S0-BRX-694

Planting A Future

PROFILES FROM OREGON'S NEW FARM MOVEMENT

BY JOHN CLARK VINCENT

Foreword by Katherine Deumling

© 2014 John Clark Vincent

All rights reserved.

Design and photos by Lisa D. Holmes (Yulan Studio, yulanstudio.com)

Published in Portland, Oregon, by Yulan Studio, Inc.

Printed in the United States.

First edition

ISBN 978-0-9915382-1-8

CONTENTS

CONTENTS (CONTINUED)

ACKNOWLEDGEMENTS

A book like this would not have been possible without the people in it who took the time to share their stories and agree to be profiled. So to all of the farmers and folks supporting those farmers, I say thank you. Thanks for talking to me, and more importantly, thanks for the work you are doing. Without you, and those like you, truly good, nutritious food, and healthy, robust soils would be a thing of the past.

I also would like to thank everyone who has assisted with the creation of this book. Thank you to my friend Jane Pellicciotto, the Slow Food Portland board member who introduced me to my book's Foreword author, Katherine Deumling, and thank you, Katherine, for your effort and support of America's new farm movement. Special thanks to Amber Holland and Anna Curtin of the Portland Farmers Market for sharing food-to-consumer knowledge. Thank you to Katie Swanson, who hopes to become a farmer, for helping me understand some of the challenges of entering the field. And thanks to all my friends who have read and provided feedback on various drafts during the writing process – Waka Takahashi Brown, Jane Pellicciotto, Matt Merenda, Pam Zipfel, Linda Bybee-Kapfer, and Mary Ann Whitney-Hall.

It's important to acknowledge some of the organizations and online resources that contributed to my understanding of the U.S. food system, where it's been, where it is, and where it's headed. Organizations like Friends of Family Farmers, Organic Seed Alliance, Organic Farming Research Foundation, World Wide Opportunities on Organic Farms (WWOOF), Oregon Tilth, Oregon State University Small Farms Program, Biodynamic Association, Rodale Institute, and so many more, especially the United Nations Conference on Trade and Development's "Trade and Environment Report 2013 – Wake up before it is too late: Make agriculture truly sustainable now for food security in a changing climate."

I also would like to acknowledge a Johns Hopkins University class I took through Coursera, which provided a lot of foundational knowledge – "An Introduction to the U.S. Food System: Perspectives from Public Health." I found several papers and video presentations by Alan M. Kapuler, PhD, that were related to plant biology and public domain plant breeding to be very enlightening and inspirational. And I found the writings of Dr. Vandana Shiva to be informative and persuasive.

Finally, and above all, I am grateful to my wife, business partner, graphic designer, and photographer, Lisa D. Holmes… who helped express the idea behind this book, and who makes everything else in my life more meaningful, easier, and better.

John Clark Vincent

FOREWORD

I cook, for my family and for a living. Everyday I use vegetables grown by the likes of those featured in this book. I think of their farms, the farmworkers who picked and washed the produce. When it rains or doesn't rain or freezes or storms I think about those who grow our food. When the last of the peppers go and the first of the winter squash arrive I think about a season's worth of sunshine captured in that winter squash for us to enjoy in the cooler months.

There are many critical components of the food/farm movement to explore. I teach people how to cook and find the simplicity, joy and pleasure in local produce. Others work on food access and equity issues; still others agitate for better working conditions and wages for farm workers. John Clark Vincent tells the stories of individuals who farm, breed seeds, or support farmers in myriad ways. Whether as farmers, ranchers, seed breeders, or institutional supporters and advocates, these people are thinking both globally and locally and are applying their skill and creativity and every last resource, to grow and raise food for all of us.

Slow Food founder and leader Carlo Petrini talks about elevating farmers – thinking of them as the "professors and heroes" of our society – those who sustain us and who's skill keeps culture and soil intact. That skill and perseverance and intelligence is evident in the profiles in this book. Whether it's Daniel O'Malley's management intensive grazing practice (Sweet Home Meats) or Andrew Still and Sarah Kleeger's (Adaptive Seeds) work to bring the value back to open pollinated and heritage seeds – no small feat in a world of hybrid and terminator seeds – these people are doggedly pursuing their dreams.

I work with farmers, and some of them are the farmers in this book. Years ago I sold five pound cloth bags of Greenwillow Grains flour to friends out of my living room before there was any Portland distribution. I was so excited to find local grains and beans – through the Southern Willamette Valley Bean and Grain Project of which Greenwillow is a part. I write recipes for farmers, their CSA members and market customers. I teach people how to cook, which entails sharing a lot of food with people and telling stories about those who grow it. It never gets old. I am lucky to get to work with and support and know about the people in this book and dozens like them. I see people's eyes light up because of the flavor of the food I prepare and then see them make the connection between flavor, place, community and health. I see them remember flavors from their youth, from many different cultures and places.

I walk the fields with some of these farmers and hear about a new varietal (likely developed by Wild Garden Seed or Adaptive Seeds) or a particularly prolific crop of tomatoes. And just as often we puzzle over the ongoing economic challenges many of them face, the challenge of paying farm workers or themselves a fair wage and providing benefits if possible, and charging the real cost of producing the food – the result of the industrialization of agriculture and the drive toward cheap food John describes in his introduction. What is the right scale? Does one grow and diversify, like Minto Island Growers or does one simplify and cut the acreage in half like Barking Moon Farm, in order to be viable and sustainable, personally and professionally?

Planting A Future is about people choosing to live and work in a certain way. It's a story of people who embody and strive for sovereignty – food sovereignty. Those featured express a holistic approach to life, land, product and community. How food is produced, by whom and for what price it is sold is all part of the equation. Greenwillow Grain's story about personal health crises leading to dramatic change in growing practices and overall approach is a perfect example. Many profiled are burdened with debt and face serious threats to their future as small farmers.

All of those featured in this important book are both eminently practical and philosophical. I imagine it's their closeness and love of the land, the rhythms and peculiarities of the seasons and a changing climate, the animals, and being in the fundamental business of growing food. They know how to fix old machines, diversify their income streams, and when to move cattle to the next pasture. And yet the vulnerability of being at the mercy of the seasons and a fickle market place and a culture of scale makes them both more in control of their lives and their food source than most and yet economically so very vulnerable. What does this say about what we have done to food and food production over the last several generations?

These stories ultimately convey the complexity of farming and ranching in a holistic, sustainable way – why we need organizations like Friends of Family Farmers, Adelante Mujeres and OSU Center for Small Farms and Community Food Systems. It's a valuable book and one that calls us to action – to become co-producers, like Slow Food's Carlo Petrini likes to say, to not just consume but think of ourselves as an important part of the production itself.

Katherine Deumling
Owner, Cook With What You Have
Chair Emeritus, Slow Food USA

OVERVIEW MAP

1. Abundant Fields Farm
2. Adaptive Seeds
3. Adelante Mujeres
4. Barking Moon Farm
5. Deck Family Farm
6. Fiddlehead Farm
7. Friends of Family Farmers
8. Greenwillow Grains
9. Lonely Lane Farm
10. Minto Island Growers
11. Naomi's Organic Farm Supply
12. OSU Center for Small Farms and Community Food Systems
13. Peace Seedlings
14. Sweet Home Farms Meats
15. Vibrant Valley Farm
16. Whistling Duck Farm
17. Winter Green Farm
18. Your Backyard Farmer

This is a book about farmers. People who, day by day and season by season, give shape to the vision and the hope for a better world.

Whether you know them or not, the farmers in this book are your friends. They care about your health, your families, and your communities, even if they don't know you personally. And it's important that they do. Because in many ways, these farmers hold our future in their hands.

These farmers are not the corporate, mass-production food manufacturers who fill our supermarkets with unhealthy junk. They're entirely different. They live by a different set of rules, with rule number one being – take care of the soil. Be good stewards. Make it healthy so it will live to produce wholesome food for generations to come.

It's unfortunate that the vision of a healthy earth filled with healthy people is an "alternative" view of life and of agriculture. But perhaps, if we work with these farmers, we can begin to change that. To do that, we need to understand what these farmers are working to overcome.

FOOD IS LIFE

Here is a fact you probably know – the quality of the food we eat has a direct influence on the quality of life we lead, because our bodies are made of the food we have eaten. Literally. The plants and animals I ate yesterday are no longer plants and animals... they now exist as a part of me. They were transformed into a new state of being by the process called 'eating.'

If I eat healthy, whole food, then I will become healthy and whole. If I eat junk food or food poisoned with pesticides, then I gradually will become chronically unhealthy... physically, emotionally, and mentally.

Unfortunately, the majority of food in our grocery stores is junk food or food tainted with poisons. A high percentage of our food contains insecticides, herbicides, chemicals, and genetically engineered sweeteners. This situation exists because our country's current industrial food system is based on the use of chemicals, poisons, and other "junk" to mass produce our food. And this reality is creating health and economic challenges for all of us.

BUILDING THE FOOD MACHINE

War has played a big part in the development of our American economy. When we take part in a major war, like World War II or the Vietnam War, many businesses that make products used for war – like bombs and chemical weapons – grow with the "war economy," and they want to continue to make a lot of money after the war ends. They do this by adapting their products from a wartime application to a peacetime application.

For example, after World War II, the chemical companies that manufactured bomb-making materials figured out that those same chemicals could be used as a type of fertilizer on farm crops. And after WWII and the Vietnam War, poisons like those taken from Agent Orange could be used to kill weeds and pests on those same farm crops.

It took a lot of advertising and marketing, lobbying and deal-making to get it done, but these companies succeeded in becoming major agribusiness companies.

As these companies became increasingly involved with agriculture, they realized they could profit even more if they could control the seeds that farmers used, initially through hybridization and ultimately through genetic engineering, and the patenting of those seeds. They did an excellent job in this endeavor, because today nearly 80 percent of all seeds are either owned or controlled by about five different companies. Which means that many farmers have to buy new seeds every year rather than keep the seeds they grow. Since the beginning of agriculture, the freedom of seeds has been a foundation upon which agriculture is built. But now the corporations rather than the farmers are in control.

On the animal side of the equation, big industries figured out that if they put huge numbers of animals in very small spaces and fed them diets that made them get big and fat very quickly, they could reduce their costs and make a lot of money selling the meat. And that business approach led to what are called Concentrated Animal Feeding Operations, or CAFOs. Today, CAFOs produce from 50 to 80 percent of the meat we eat, depending on the type of animal, and most of that meat is processed by only four different companies.

Since about 1950 (following WWII), farming in the

United States has been transformed from a nationwide network of small family farms to an immense agrochemical industry (Big-Ag) controlled by a few very large companies.

SO WHAT'S THE PROBLEM?

Admittedly, the above explanation for how we got here was so simplistic that it might seem like there should be a simple solution. But how we got here was actually a complex process that occurred over several generations. And the solution will probably take at least that long. But let's begin by looking at why all of this is a problem.

Life on earth is completely interconnected. Everything which is alive is made of the same basic material, and all living things are continually interchanging that material at the cellular level. Our bodies were created by life's immense diversity and the myriad relationships within that diversity. And to function properly, our bodies need to continue to be a part of immense biodiversity.

For quite some time now, the vast majority of us have depended on a system of food creation called agriculture to get the food we eat. And the food we eat is our number one connection to the earth's biodiversity… our connection to the power of life. So agriculture – how we farm – is one of the most important things we do to keep our world healthy and to keep our bodies healthy.

The problem we face today is that the majority of the food available to us is not being produced in a healthy way.

WHAT'S UNHEALTHY ABOUT IT?

Understanding that we need a healthy diversity of life in order to be healthy ourselves, it's clear that healthy farming begins with biodiversity. So the first unhealthy aspect of modern industrial farming is that it often strives to kill pretty much everything in the field and leave only one type of plant standing. This is called monoculture, or single crop, farming.

In a healthy farm field, there are gazillions of organisms living in the soil. There also are many plants growing from the soil and decomposing within the soil. There is water, or moisture in the soil that the plants drink. And there are many animals, most of them insects, living among the plants. This is what a healthy, biodiverse field looks like.

Today's industrial monoculture system is based on destroying this diversity. Over a very large area, the seeds of only one type of plant (like corn or soybeans) are planted and

sprout. And then to make sure no other plants are able to compete with these corn sprouts, poisons are sprayed on the field to kill every type of plant except the corn. To make sure no insects can eat the corn, more poisons are sprayed on the field that kill all the insects. These poisons seep into the soil and also kill many of the organisms in the soil. So basically at this point, the corn is growing in a field that is essentially barren.

The system keeps the corn alive by feeding it chemical fertilizers and giving it lots of water to drink by irrigating the field. But how is the corn not killed by the poisons that killed the other plants? Scientists hired by the chemical companies placed a gene inside the corn seeds that keep the corn alive when it's sprayed with poison. And in some cases, the corn itself becomes a poison or makes its own poison to kill any insects that try to eat it.

That's how it works with the plants produced in our current food system, and it's really no better with the animals.

All animals, including us, are healthiest when allowed to eat healthy, natural diets and exercise outdoors in healthy spaces. But in our industrial food production system, almost all the animals we eat are packed into tight quarters so they can't move around freely. They are fed the type of corn and soybeans described earlier, even though in some cases these animals are meant to eat grass and can't properly digest corn and soybeans.

Between their filthy, confined living conditions and their unhealthy diets, it's easy to understand why many of these animals would get sick. To prevent that, they are given antibiotics all the time, whether they are sick or not. As you know, antibiotics are designed to kill all the bacteria in our systems, both bad and good. Plus, the antibiotics make the animals gain weight faster. And even though it's unhealthy weight, more weight brings in more money, so the people in control view it as a good thing.

HOW DOES THIS AFFECT OUR LIVES?

When farm animals are given antibiotics all the time, bacteria in the animals begin to develop immunity to those antibiotics. Over time, the antibiotics – the medicine we depend on to combat serious infections – become less effective for their intended purpose because they have been over used. As a result, we are currently in the process of losing one of our primary healthcare tools.

In the fields, the poisons that are sprayed on everything don't just magically disappear. They seep into the soil. They get into our water system. They kill insects of all types, even insects like bees that we depend on for our food supply. They are found in the air we breathe. They also accumulate inside the farm animals that eat the poisoned corn and soy. And they gradually accumulate inside us. Recent research has shown that one of the most popular pesticides is even found in human breast milk. These are the effects of the poisons, but there are other industrial effects that aren't quite as easy to see.

One of the reasons our farming system has become this way is because many people thought agriculture should be turned into a mass-production industry and made to work the same way other manufacturing industries work. If the farms got larger, they would benefit from economies of scale. If mass-producing cars works so well, the same ideas should also work with mass-producing cows. Or pigs. Or chickens. Or soybeans. Or wheat. Or corn.

Mass production is built on specialization and controlling the production environment. So when we look at the large fields with only one type of seed growing, or the massive feeding operations where cattle or chickens are confined and pumped full of grain and antibiotics, it's easy to compare this to a factory where food is made. Life and health have been taken out of the equation, and our food is simply being manufactured. Mass-produced.

Once the food has been mass-produced, it goes to other factories that turn it into the food we eat and put it into packages that are attractive to us. By this time, all of that unhealthy food doesn't taste very good anymore, so more chemicals and sweeteners are added to it that will fool our tastebuds and make our bodies believe they are eating something very good and tasty. And eventually, we become addicted to the drug-like effects of these products. We become the consumers that every mass-production system depends on.

WHO IS AT FAULT?

When we learn that something about our lives, or the lives of our children, is unhealthy, we have a natural and normal tendency to want to blame someone. But in this case, we only have ourselves to blame for the problems with our food system.

Over time, as a society, we became less involved with

our food. The idea behind the American Dream became more materially focused. It required more time and created more stress. With less time, we began to search for ways to make life more efficient. Our lives began to look a whole lot more like they, too, were being manufactured. Conveniences and short-cuts increased in value, and the food industry was right there every step of the way to help make that possible.

From fast food hamburgers that never decompose to boxes of food-like substances you only have to heat in a microwave oven, our food ceased to look – or taste – like real food.

It's easy to understand how a young couple with two jobs, kids, car payments, college debt, and a mortgage would become detached from the process of growing, cooking, and eating healthy food. Is it their fault? Yes, in a way… but many pressures and cultural influences made it almost impossible to avoid – I should know. I grew up on a farm, and yet this happened to me, even though I had an awareness of what real food is. I found myself living in a city, working in a corporation, and eating dinners that only required opening a box.

But it also is the fault of companies who knowingly deceive people and promote their addictions. Tobacco companies did this for a long, long time. They used science to do it. Scientists hired by tobacco companies provided studies that proved tobacco was safe… until years later when independent scientists proved that it wasn't. A great deal of what appears to be tobacco science is being conducted today with industrial food processes, and we need independent scientists to step forward to help determine the extent to which our agrochemical food machine is unhealthy.

At the same time, we need to keep in mind that bioscience is not bad in and of itself. Biological, chemical, and physical sciences help us understand many important things, and help us solve many problems. In fact, to a large degree it's science that is showing us how bad our current food system is. And science will help us correct it. But that science needs to be done by honest, independent scientists who are not on the payroll of the industrial food corporations. And while that scientific work is being done, we need to take action of our own.

The first thing we need to do is accept responsibility for our food, and not waste time and energy blaming others. Because all of those 'others' are just people like us. People who want a good, healthy, happy life for themselves and their families.

WHAT CAN WE DO?

There are many people in this country who, right now, are working hard to correct our food system. They are working as organic and sustainable farmers or they are working to help and support these farmers. They are not part of the agrochemical or industrial food machine. They are creating a new food system, which is sometimes called the *new farm movement*.

Many of these new farmers are idealistic people who believe it is important to do good in the world. They are independent and like being their own bosses. They hate sitting at a desk all day and prefer working outside in all types of weather. Some farmers enjoy digging in the dirt, planting seeds, and watching those seeds grow into healthy food. Others have an affinity for animals and raise them humanely, giving them the opportunity to live the type of lives they would naturally live. And everyone enjoys sharing the food they produce with the people who eat it.

That's what we can do. We can buy and eat their food. And we also can work to make sure everyone in our communities has access to this wholesome and nutritious food. Because healthy food is a requirement for a healthy, balanced life.

Another thing we can do is join them. New farmers are needed everywhere in our country. Even in our cities. Regardless of your age, you are never too old to produce food by farming or gardening. Neither are you too young.

At the moment, many of these farmers are struggling to keep their operations running. And that's why they need our help. As the size of the community buying their food grows, so will their ability to produce more food and create more jobs for others in the food system.

WHO ARE THESE FARMERS?

The new farm movement is springing up all over the country, and Oregon is one of its strongholds. The efforts here to develop an organic and sustainable farming system began in the late seventies and early eighties. They were efforts fueled partly by the "back-to-the-land" movement that started with the hippies, Vietnam vets, dropouts, environmental activists, and free thinkers of that era. Homesteading was

common, as were food co-ops, and dog-eared issues of publications like *Organic Gardening*, *Mother Earth News*, and *Small Farmers Journal*.

Many homesteads, intentional communities, and farms from that time no longer exist, but some took root and began to flourish. In Oregon that included farms like Gathering Together Farm and Winter Green Farm, which today are some of the movement's larger, more successful enterprises. It also included early cooperatives like Organically Grown Co-op, which now is Organically Grown Company, one of the largest organic food distributors in the Pacific Northwest.

There were the seed pioneers, like Dr. Alan Kapuler of Peace Seeds and Frank Morton of Wild Garden Seeds. These two visionaries saw the coming problems with seed patents and the corporate goal of seed control, and worked to strengthen open-pollinated seed varieties and create public domain seeds that would remain free.

Over time, farmers markets began to spring up. Community Supported Agriculture (CSA) agreements between farmers and consumers came into being. Urban farms like 47th Street Farm and Zenger Farm in Portland began training new farmers and offering educational programs to children so the newest generations of consumers in America would know what real food is and where it comes from.

There are so many wonderful farmers and organizations in Oregon's new farm movement who are not included in this book's collection of profiles. But the people who are included are meant to serve as representatives of the larger community. They range from the pioneers at Winter Green Farm to a novice farmer just getting started at an incubator farm, and everything in between.

Some of them have come to healthy farming out of moral and spiritual motivations, while others simply want to find a way to earn a living that allows them to live and work outdoors and be in charge of their own lives. And a couple of profiles feature conventional, multi-generational family farms that have made a shift to embrace the ideas and ideals of the new farm movement.

Bottom line… this book presents a cross section of farmers that is meant to represent Oregon's new farm movement. A movement that is planting a future we all can look forward to.

Winter Green Farm

Originally founded as a homestead in 1980 by Jack Gray and Mary Jo Wade, Winter Green Farm has grown to be a successful biodynamic farm in Oregon's southern Willamette Valley.

Jack Gray, Mary Jo Wade, Wali Via, and Jabrila Via came together at Winter Green Farm the way tributaries meet and combine on their way to something larger than themselves. In doing so, they helped cut a path that many of today's new farmers are stepping into. There are a lot of ways to tell this story, but I think I'll follow the sun west and begin back east in Atlanta, Georgia.

Wali Via was looking for something when, as a teenager, he ran away from his Georgia home. A lot of people were looking during the late sixties and early seventies. But regardless of the inciting incident, his search began in an intentional community (aka commune) in the Georgia countryside, and it was there he discovered an activity that would become the central element in his journey. That activity was organic farming.

Wali spent ten months discovering agriculture in that Georgia community before resuming his travels and ultimately settling in another intentional community near Deadwood, Oregon. The eight and a half years he lived and worked in the Deadwood community brought to fruition all of the seeds he had carried with him from his Georgia roots. He began to understand the nature of biodiversity and the rudiments of biodynamic farming. He also met the love of his life, Jabrila. And together, they had a child and began a family of their own.

It takes very little time when talking with Jabrila Via to know she is a perfect match for Wali. A fellow seeker from Menlo Park, California, Jabrila embodies the metaphysical influences of people who cherish the spiritual connectedness of life and wish to share the beauty they see. But in Jabrila's case, metaphysics should not be confused with something overly ethereal or anything other than pretty damn strong. She has earned her position in Oregon's organic farming community with a lot of work over a lot of years.

Not long after the birth of their first child, Wali and Jabrila left their intentional community to create a homestead of their own several miles down the road on a worn out farm they were able to lease from a friend. They both talk fondly of the experience, but neither mince words about what kind of a struggle it was. With virtually no money, Wali was forced to find work where he could, either in the woods or occasionally fighting fires. Which left Jabrila home alone with first one, then two little girls.

"I'd get up in the morning with a three-year-old, a new baby, and a cow to milk. My three-year-old learned to milk a quarter by herself. She was fantastic and she'd just sit there and do it. Then I'd milk the rest of the cow while I nursed the baby, and of course the cat and the calf wanted milk, so it was a full experience. There were greenhouses to water. Starts and transplants. Things had to happen and it was me and the girls, so we'd make games of it. The big projects waited for Wali to come home."

It's hard to make progress when there's too much to do, and by 1985 Wali and Jabrila knew they were at a decision point. Either they were going to spend the rest of their lives struggling to make something of the rundown farm they were

The vibrant mix of colors in this field of cruciferous vegetables reflect the health and vibrancy of the soil that supports it. As a biodynamic farm, Winter Green Farm creates its own compost to enrich the soils. The farm managers also practice crop rotations and maintain buffer zones around fields that are home to a wide variety of native plants. These zones provide a home to the many beneficial insects that help keep the kale and other crops healthy.

attempting to rebuild, because it was going to be slow going, or they could look for another situation that might lead to more opportunities for them and their children. That led them to pay a visit to Jack Gray and Mary Jo Wade, another young farm couple they had met during the original formation meetings of Oregon Tilth.

Jack and Mary Jo had their own homestead near Noti, Oregon, which has grown to become the Winter Green Farm we know today. But when it was founded in 1980, it was just two idealistic 20-somethings chasing a dream they were struggling to realize. Regardless, they weren't about to give up on it… that wasn't in their makeup.

If one imagines what a wise and experienced farmer might be like, I suspect they'd come up with someone akin to Jack

Gray. Jack's journey to an Oregon homestead was nothing like Wali's, but it was every bit as interesting.

Jack was born and raised in Portland, son of a businessman who happened to own a ranch in eastern Oregon, and that's where Jack "was involved for a chunk of time from high school on." It's where he loved to be. Outdoors. Riding horses and raising cattle. Who would have guessed that he'd travel to Middletown, Connecticut, to attend Wesleyan University and study geology. But I'm sure he's glad he did, because that's where he met his very own Jersey girl.

When Mary Jo met Jack, she knew nothing of farming beyond the garden her father grew in their backyard, but she was ready to follow Jack west when college came to an end. "I didn't mind coming to the West Coast. I'd finished my economics degree, and I was ready to go somewhere. My boyfriend was out here so I came out here. I worked in Portland the first couple years while Jack worked on the ranch in eastern Oregon. As far as homesteading went, I was as idealistic as everyone else. It turned out to be a pretty fun adventure. And I guess it worked out because we're the only people I know who have had the same address since 1980."

Before launching their homesteading plans, Jack took a job as business manager of a new magazine called *Small Farmers Journal* which actually had its own small farm. Jack and Mary Jo moved onto the Journal farm and if Jack had any reservations about farming prior to that, the magazine erased them. Readers of the Journal can vouch for the fount of agricultural romance packed in each issue. The Journal even had it's own draft horses. At that time, the Journal was based in Junction City, Oregon, so they began looking for their own land in that area and settled at their current place near Noti.

Jack and Mary Jo's homestead had a stronger financial foundation than Wali and Jabrila's, but homesteading is homesteading, and regardless of what you start with, a steady source of income is necessary to sustain it.

"As we got started we were going to raise raspberries and be a dairy," said Mary Jo. "Maybe sell cheese. And do it all with draft horses. We were pretty young. We had a lot of ideas."

"We were really naive about how to make it all work," added Jack. "We were pretty grounded in the environmental movement and we had a lot of things in our head. But within a couple of years we got rid of the draft horses. When we looked at the prospect of becoming a viable farm and actually making some money, we realized we had to do something different. As we looked around, we noticed that a new environmentally friendly, alternative form of agriculture was just starting to bud a little bit, so we tried to get involved with that."

So they began attending the early organizing meetings of Willamette Valley Tilth (which would evolve to become Oregon Tilth). They also participated with another local effort called the Organically Grown Co-op (now Organically Grown Company) which was in the process of kicking off. As they ventured out, what they discovered was that many things involved with food and agriculture were starting to change.

Diverse activity everywhere. So many possibilities.

Which brings us back to that meeting with Wali and Jabrila. Two couples, both convinced they could accomplish more by working together with someone else than they could do alone. There was no intention to do a deal when they all sat down. They were just two families getting together to share ideas and maybe ask for a little advice.

Wali recalls, "We just came over to talk to them and after awhile they asked if we would consider working here, and we thought that sounded like a good opportunity so we said okay. Then we worked here and we did a contract with them that first year for vegetable starts. That's my recollection of it, and that was 1985."

It wasn't all roses though. After their years farming in the Deadwood community, and another five years working to build a life on their own farm, Jabrila didn't want to leave it all behind. Turns out there was a silver lining waiting to be discovered.

"I believe the thing that really changed us over to doing something different was that Wali never got to see his kids," says Jabrila. "I didn't want to come here at first, although I'm very happy here and it's my life, but I didn't want to leave our farm in Deadwood. I loved our life there. But when the girls woke up that first morning here and they asked 'where's Dad?' I said 'look out the window.' And he was out there harvesting cauliflower and they were like, 'is this where he works?' and they ran out and saw him, and I was like, okay, okay, I'm over myself."

Thus began a partnership that would yield one of the best loved organic farms in Oregon, as well as one of the few biodynamic farms of any size. That's not surprising when considering the spiritual and idealistic roots these four farmers share. For those who aren't familiar with biodynamic practices, it kind of takes an organic commitment and kicks it up a few notches.

A biodynamic farm is viewed as one holistic organism. To the extent possible, no external inputs are brought in. Instead, all inputs such as fertilizer and compost are generated on the farm, which means that both plants and animals are raised in harmony, and all of the farm's processes are intricately connected. A healthy piece of earth is home to a diversity of

In addition to its vegetable sales to CSA and farmers market customers, Winter Green Farm grows a large amount of a root crop known as burdock for the wholesale market. Burdock is called 'gobo' in Japanese and is used extensively in Japanese-style cuisine.

plants, both those we call crops and those we call weeds. Both insect pests and insect beneficials have their role to play. The soil needs to eat and drink and breathe to become supple and strong.

There's a complex science behind biodynamic agriculture that was founded by a man named Rudolf Steiner. But if you just sit back and think about it, a biodynamic farm is pretty darn close to the classic, old-time family farm with cows and pigs and chickens and goats and grains and vegetables and fruit and pastures and all the rest of it. Throughout history those farms managed to produce food as self-contained operations that were handed down through the generations. They didn't depend on chemicals, poisons, or even organic imports being brought onto the farm the way modern industrial agriculture does. They depended on healthy soil and diverse natural ecosystems. And anytime those principles were cast aside, the land perished and became barren.

That's one of the truly beautiful things about Winter Green Farm. It has taken an idealized notion and, through science, management, marketing, and dedication, has translated that notion into a commercially viable agricultural reality.

Another important part of the Winter Green Farm story is that it has served as a training ground for hundreds of future farmers and contributors to the organic agriculture industry. Some for just a season or two, but others stay for many years. Two of those long-time employees recently became the

third couple to join the farm's ownership group. After fifteen years of commitment to Winter Green Farm, Chris and Shannon Overbaugh became co-owners in 2009.

Chris and Shannon exemplify the approach that Jack believes works best for people who want to become farmers. "You really have to work for other farmers first," he explains. "Learn and see what it takes, and then build a plan from that. A lot of people are homesteading these days, and homesteading is great, but that's not really the best way to break in to farming anymore. Also, I think there's a lot of potential for cooperative farms, but the main thing is that there's just a lot more to learn about organics now than there used to be. Less margin for error. More professionalism is required. It's one thing to grow it, but it's another one to produce a product that you're able to consistently sell.

"And I suppose that it's important to keep an open mind. In the long run I think agriculture will just keep evolving. Commercial agriculture has definitely moved in the direction of organic. And unfortunately, organic has moved a bit in the commercial direction. To a certain degree, it may need to. But the question of scale is a central issue. Can you really scale up organic and have it work in the same way... the holistic way it should work. Large farms do have some incredible advantages economically, but in terms of where things should be going... I'm a firm believer in local, and that's not going to change."

(Clockwise from top left) Basil is grown to produce pesto. A magnificent, late-summer field of carrots. And part of Winter Green Farm's herd of cattle.

ABOUT

WINTER GREEN FARM
Jack Gray & Mary Jo Wade; Wali & Jabrila Via; Chris & Shannon Overbaugh
Noti, Oregon

Winter Green Farm is an organic and biodynamic farm located in the foothills of the Coast Range Mountains west of Eugene, Oregon near the small community of Noti. The farm was founded in 1980 by Jack Gray and Mary Jo Wade. In 1985 Wali and Jabrila Via joined the farm's ownership group, and in 2009, that group was expanded to six with the inclusion of longtime employees Chris and Shannon Overbaugh.

Winter Green's 170 acres provide mixed vegetables and berries for a large-scale CSA operation and direct sales at farmers' markets, as well as basil and burdock for wholesale accounts and a pesto business. About 90 acres of the farm are dedicated to cattle, pasture, and hay, and the remaining acreage consists of oak groves, riparian areas, and wetlands for wildlife habitat.

For more information about Winter Green Farm, please visit their website at wintergreenfarm.com.

Fiddlehead Farm

Fiddlehead Farm is nestled in a picturesque clearing in Corbett, Oregon, where the northern Willamette Valley begins its climb to meet Mt. Hood in the Cascade Mountains.

Rowan Steele is self-possessed and confident. Quick to smile. There's a natural energy in the way he moves around his farm, sharing his crop rotation strategy, pointing out the worst erosion spots and explaining how they were solved… if they were solved… the coming rains will provide that bit of wisdom. But he believes his new drainage system will work, and I do, too. Because it's hard to not have faith in this young man.

Katie Coppoletta is solid. Grounded. Maybe a little cautious at first, but you can watch the trust settle in and her easy country manner take over. She's not a dreamer in the same way Rowan is but she knows what she wants and she's willing to put in the work. And anyone who's spent much time on a farm would swear she was born to it. That's not the case. Not for either of them.

But when Rowan and Katie first began experimenting with agriculture, both immediately felt drawn to it, although they'll be the first to admit they had no idea what it would take or how far they would need to journey to become real farmers.

Farming is a profession after all. And like any profession, one does not rise to proficiency – much less mastery – without years of training, practice, and getting up every morning and going to work.

That's what Rowan and Katie have given to farming. In return they have received a way of living that suits them perfectly. But is it a lifestyle they would recommend to others?

Katie puts it this way… "Farming can be really rewarding, but it takes a certain type of person to enjoy it because there are so many challenges. If you're a person who can deal with that, it's a wonderful lifestyle. Much of the time it doesn't even feel like work."

She continues by pointing out that she doesn't mean there's only one way to go, or that a person has to fit a specific definition. It's simply a matter of finding the path that's right for you.

"For me, the key is direct marketing. That's probably the only way I could be a farmer. It wouldn't be very fun if we were separate from the people we are selling to. That's really important for me… it's like the end result, you know. Knowing people who appreciate what you're offering them makes it all worth it. If I don't have that it's easy to get resentful."

Rowan explains, "You get that way because you lose sight of the bigger picture and get stuck in the day to day operations. You're mentally and physically exhausted, and it feels like you're not getting anywhere. That happens a lot. So you have to step back and take a breath.

"One of my favorite remedies is… on an autumn day like today, end of the season, it's not super hectic, come sunset, Katie and I just go to the top of the hill and sit and look at everything we've accomplished. It's super rewarding to see that we're making progress on something we've invested so much of ourselves in."

Like many members of Oregon's sustainable farming community, neither Katie nor Rowan began their search for a career thinking they would become farmers. She was an

A new barn anchors the farm Rowan Steele and Katie Coppoletta are building. It serves many roles, like providing the washing station and holding the coolers, and in autumn is filled with garlic. The peppers in this photo are grown for Hot Winter Hot Sauce, a local Portland hot sauce company.

international studies major, a field that introduced her to global food systems. And although Rowan grew up in the central valley of California, a sort of agricultural mecca, farming didn't occur to him until he began studying geography and got exposed to different ideas and different ways of producing food.

Katie first became interested in farming at the same time she and Rowan were students at Humboldt State University in Arcata, California... but it probably didn't happen the way you might be thinking. Even though they went to the same school, they didn't know each other until they met while studying abroad in Mexico as part of a sustainable development program.

It was there, in 2005, that Katie found her future husband and got her first taste of agriculture... "I helped at-risk kids develop a garden, even though I knew nothing about gardening at the time. I partnered with a friend who had some

farm background, and I just jumped into it and double dug this whole grassy area at the orphanage. I put my heart into that project, and it just felt really good."

After connecting in Mexico, Rowan and Katie returned to Humboldt where Katie continued working with at-risk youth while Rowan began studying sustainable agriculture and volunteering at a friend's farm. But Katie couldn't resist the draw of farming... "I was living vicariously through Rowan's volunteer work at the farm until I couldn't stand it any more and started volunteering with him. But they didn't really need me, so I ended up volunteering for a different farmer down the road who was kind of struggling. I essentially became an unpaid employee on that farm because I wouldn't accept his money. I did that for a year and a half, which gave me some really good experience."

During that period, Rowan rose from farm volunteer to co-manager of a sizable Community Supported Agriculture

(CSA) operation at Arcata Educational Farm (AEF) – an informal incubator program for new farmers.

"Taking over a 60-person CSA was a challenge and we definitely made some mistakes, but I learned a lot. After that experience, I knew I'd always be working with the earth… working with it in what I deemed a meaningful way. I also knew that I wasn't really cut out to run a CSA, but I am glad I did it because it put me on the path I was meant to be on."

Strong words, spoken from the heart, and Rowan waxes poetic when he tries to define that path. "What I realized from my brief two-year stint at the AEF was that farming was extremely gratifying and fulfilled me in a way that I'd never experienced.

"It was productive work that connected me to everything that really mattered to me: community, mental and physical stimulation, delicious and nourishing food, problem solving, creativity, art, system design, skill development, simplicity. It crosses cultures, and is something that connects me to previous generations. Practicing agriculture is participating in a truly universal human experience. It is an activity I am simply programmed to do."

The experience he had at the AEF was Rowan's introduction to a farming lifestyle and, even at that early stage, he knew it would be his guiding force leading forward.

When their work in Arcata ended, Katie and Rowan did some traveling, including some international farm interning called WWOOFing (World Wide Opportunities on Organic Farms). They both believe it was then – when they were working on a farm in Argentina – that their thinking solidified around the idea of wanting to start their own farm.

After returning to the U.S., Rowan and Katie moved to Portland, Oregon, where Rowan began graduate school at Portland State University while Katie found work at Portland's highly respected 47th Avenue Farm.

As Katie gained greater experience with a large commercial agricultural operation, Rowan worked on a graduate study program that would benefit not only his own farming future, but the future of many other beginning farmers.

"Much of my time in grad school was spent developing a proposed farm incubator program for the Portland area.

I wanted to create a mechanism to help other people get started in farming while promoting scales of agriculture that I believe are beneficial to a community. But I also focused on the conservation piece… making sure that conservation agricultural practices and good stewardship are used. I went to grad school with the intent of doing something with community agriculture, and this incubator study seemed the perfect way to do that."

Eventually Rowan's research led to his current position as incubator farm manager for the East Multnomah Soil and Water Conservation District – a position that helps support the legacy he and Katie are building at Fiddlehead Farm. But that legacy doesn't come easy. Several years in and benefiting from Katie's retired parents now living on the property and sharing the financial burden, farming remains a genuine challenge. And the biggest hurdles always are the same… dealing with the abundance of work and the scarcity of money.

Rowan believes there's money to be made. "You can make a living… there's money to be made, but it just takes a long time. Frank Morton [of Wild Garden Seeds] said at a local conference that eventually the poverty starts to go away. You have to make it work with other things in your life. If Katie wasn't committed to being a full-time farmer, we couldn't pull this off. I'm just not around enough. And then if we didn't have my additional income we wouldn't be building the barns and greenhouses. But we've found our balance. I get enough hands-in-the-dirt time to keep myself satisfied, and we're able to get enough help to keep her sane. That's how we've made it work. I think that each farmer and each farm needs to find their own way of making the operation work."

For Katie and Rowan or any other young farmers, one essential element of making ends meet and finding success is integrating yourself into the local farming community.

Katie points out that beginning as a farmer means learning to be alone a great deal. "You have to be okay with a certain degree of isolation. Eventually you'll probably add some employees or interns, but that takes time. I believe that one reason a lot of farmers don't make it is because they feel like they're alone and not part of a larger network of like-minded people."

Rowan agrees, "We've been able to be successful as a

Kale is another Fiddlehead Farm specialty, much of which they sell to local Portland grocery chain New Seasons Markets.

farm because we came into it with a strong community. We've had no problem leaning on that community or being a part of that community to support other people when those times come. We certainly recognize that we're not doing this in a vacuum."

Being located close to the Portland metro area helps Rowan and Katie feel attached, which they see as an advantage that farmers living far from a population center may not benefit from. Also, when a person who is trying to farm sustainably is surrounded by neighbors who don't share the same approach or goals, it's easy to feel like an island.

Katie sums up community this way… "We know we're all here for each other. We talk on the phone. We sell each other stuff. We can feel the community. When you find people you like and respect and they like you and have the same vision… you hold onto those people."

Armed with experience, community, multi-generational support, established markets, personal commitment, and land they own, it appears that Katie and Rowan have put all the pieces together to have successful careers as farmers. But even with this degree of preparation, there are no guarantees for small family farms. In fact, the cards are stacked against them.

Enormous quantities of tax dollars go each year to support our country's industrial food system, in spite of the fact that it's been proven repeatedly that the way we grow and consume food in this country will eventually bankrupt us… both financially and environmentally.

But Rowan believes people are beginning to recognize the limitations of the current food system. "From a cultural standpoint, from an economic standpoint, from an environmental standpoint, I think people are beginning to see that things have to change. Or maybe I'm being naive. Maybe that's just the Portland bubble talking, while most people remain completely ignorant about how it's impacting their lives.

"To be honest, on this subject I'm kind of a doom and gloom guy, and I think it's going to take a big market shift. Some sort of shock to the system that will ultimately lead people out of it.

"Food is so complicated. As producers I think we try and do as much as possible… we stick the seed in the ground, we grow our crop, we harvest it, and we bring it to market. That's our piece, but it's so much bigger than us. It's so far out of our control that I think if you tried to look at it from a big scale all the time and tried to base your operation off of exclusively your own values, you'd just be swallowed up by it. It's too big and too complex. And as long as corporations can invest money in our elections, we probably aren't even going to get labeling of genetically modified organisms on food packages.

"But we can't let that stop us. While we're farming this land, we have a responsibility to do it in a sustainable way. Good stewardship of the soil and the water is about more than just us. Will our child want to farm this land? Who knows. But if she doesn't, we both hope that someone else will. Our job is to create a farm that will sustain life for many generations to come. And that's what we intend to do." 🍴

(Clockwise from top left) Garlic hanging to dry from barn rafters. Kale in the foreground, backed by one of the farm's greenhouses. Seed garlic ready for planting.

ABOUT

FIDDLEHEAD FARM
Katie Coppoletta & Rowan Steele
Corbett, Oregon

Fiddlehead Farm is a small family farm located in Corbett, Oregon. Katie Coppoletta and Rowan Steele started their farm business on leased land in 2009 under the name Greenthumb Garlic, then founded Fiddlehead Farm at their current location in 2011 with the help of Katie's parents, Kathy and Joe, who also live on the property. In 2013, the birth of Ella Luna Coppoletta Steele brought the number of generations on the farm to three.

Fiddlehead focuses on the production of organic vegetables for both direct sales to consumers through farmers markets and some commercial accounts for specialty crops like garlic and kale. The farmers emphasize the use of organic methods, sound conservation practices and good stewardship of the soil, with the long-term goal of creating a farm capable of sustaining healthy and productive agriculture for many generations to come.

You can meet Katie and Rowan at Portland's Montavilla Farmers Market and read more about their farm and products on their website: fiddleheadfarmers.com

Adaptive Seeds

Adaptive Seeds, inspired by an international seed sharing project, sells public domain, open pollinated seeds. Most of its seed is adapted to the Pacific Northwest and similar season northern climates.

When first meeting Andrew Still and Sarah Kleeger, it's hard to imagine that these two young farmers spent their early years in Southern California… one in Ventura's suburbs, the other in apartments near Anaheim. Because these days they're pretty well countrified.

That's not so much a reflection of how they look, although Andrew's beard, Sarah's braided hair, and the pair's simple, well-worn clothing fit the image one might have of young organic farmers. It's more an acknowledgment of their comfort; their ease as they walk through rows of last year's vegetables in late winter… stopping to share cucumber-flavored sprigs of salad burnet, a plant they readily admit saved their ass a few times when they were running a winter CSA.

The burnet is the lone green perennial in a small plot just south of their aging farm house, which doubles as their seed warehouse. These slumbering rows of last year's garden sit beside the seedling house and adjacent to the equipment shed – home to a small tractor and bargain-priced combine which Sarah claims that Andrew loves to wrench on.

Walking east through the larger fields, they both laugh in a roll-your-eyes sort of way as they point out two completely different types of winter cabbage that were supposed to be the same variety. One of them is producing pale green savoy heads, but the other is a disintegrated mess. The seeds were purchased from different vendors, and one of them didn't get it right. All seed people make mistakes, they say, and this was clearly one of them. There's no sense that they hold any kind of grudge or

feel short-changed. They simply find the mix-up interesting.

When we reach the end of the cabbage patch, Sarah heads north with an intention she doesn't share, while Andrew turns south to check out some other thing… but after about ten steps, he turns back, sees my indecision, laughs, and tells me to come with him. A quick visual tour of the farm's 30 acres ensues as Andrew points out fields of legumes and grain that run to the end of their property, where the land rises quickly into what's left of the forest.

Earlier Sarah had said, "I just now looked up the hill, and the only big trees in sight are all laying down, so it's kind of bumming me out. It's the clear cut… they're widening it. Whatever. I'll get over it." And that statement kind of exemplifies the attitude at Adaptive Seeds. This couple just sort of rolls with it, whatever comes. Their entire agricultural existence has followed that path. The Seed Ambassadors project is a perfect example.

After spending several years working for a variety of organic farms, first in northern California and then in Oregon, Andrew and Sarah were looking for a way to combine their nascent but growing interest in seed saving with a trip to Europe… a sort of working vacation.

So they added to their personal seed collection by gathering seeds well adapted to the Pacific Northwest and took off for Europe under the banner of a program they and their friends had named The Seed Ambassadors. Meetings were set with other seed savers and organic seed organizations, and they did a lot of seed swaps. Switzerland, Germany, Russia,

Adaptive Seeds' location in the central Willamette Valley places it in the heart of one of the country's most prolific and important organic seed production regions, where maintaining seed integrity is critical for all crops, including this field of Japanese Buckwheat.

Great Britain, Italy… they made the rounds and along the way managed to forge some lasting relationships within the seed community. It was a grand adventure, and importantly, as Seed Ambassadors, Andrew and Sarah seemed to have found their calling. But in keeping with their natures, it was mostly just adapting to an opportunity that presented itself.

"It had never occurred to us that we would start a seed company until we were maybe three quarters of our way through the first seed ambassadors trip," explains Sarah. "People just kept telling us that we were going to have to start a seed company."

"They would say, 'you have hundreds of seed varieties,'" added Andrew. "You need to start a seed company.'"

Sarah laughs and continues, "And we were like, oh no… we're not capitalists. We don't want to start a business. We're not trying to make money. We're doing this for love."

Andrew admits, "We felt like the whole ambassador thing sounded like a pretty good idea… we could be travelers and go collect seeds, and then it snowballed into something way more than we ever thought it would be."

They returned from their first European journey in early 2007 and immediately began growing out some of the seed varieties they had collected, mostly thanks to the owner of the farm at which they were working who gave them a couple 200-foot beds where they could do their own thing. They also were active in organizing a seed swap in Eugene, Oregon, as well as listing a number of varieties on the Seed Savers Exchange, but they couldn't help feeling a need to reach a wider audience. Plus, they weren't quite finished with Europe yet.

Andrew explains, "We went back to Europe, to Romania initially because a Peace Corps volunteer we had met organized some seed saving workshops for us. But in all, we ended up going to about ten different countries and collecting more than 800 varieties of seeds… most of which were not available in the United States. And really, it had become an almost overwhelming burden by that point. Then we met a guy named Ben Gable, who owns Real Seeds in Wales, and he was like, 'you're going to start a seed company and it's going to be awesome. What you're going to do is take all those varieties you have, trial them and see which ones grow well, choose the best, and start selling them.'"

Sarah adds, "And he said we would double in size every year. What's amazing is that he's actually been spot on. But I remember when he first told us that, and we so unsure of ourselves. It definitely helped when we heard the same thing from others."

"Oh yeah," Andrew added, "after a couple of years of working at this, Tom Stearns of High Mowing Seeds came over and told us we were growing at the same rate he grew when he was a small seed company. It was the second round of someone saying it was going to work out. He let us know that it's going to be hard, but he helped us believe we had the momentum and the grit to do this. Of course, you're always questioning yourself when you're an entrepreneur, so it helped a lot to hear someone who knows the business tell you it's going to work out."

Sarah and Andrew eventually founded Adaptive Seeds in January 2009, and then in November of that year, they reached an agreement to lease the property at their current location, which they named Open Oak Farm. But it's hard to make a living as a start-up seed company, so Open Oak Farm began a Community Supported Agriculture (CSA) membership to generate the revenue they needed to keep their farm afloat. The CSA worked out, and though it was a lot of work, it supported the initial growth of the seed business.

Now, five years later, Andrew and Sarah believe it's time to take the plunge. They have terminated their CSA and made Adaptive Seeds their full-time commitment. But I guess the question is whether one hundred percent annual growth over a five year period is really enough to live on. It sounds good, but what does it mean in actual numbers?

Sarah admits some nerves… "this is the year that… I don't want to say make it or break it, but yeah, that's kind of what it is. And if it doesn't work it probably means an off-farm job for one of us."

Andrew remains resolute, "We're trying to figure out where the money we made in our CSA is going to come from. The good part is that we'll have that much more time to spend on the seeds, which we really needed to make the seed company flourish. And I believe if we're smart and a little bit lucky and really, really tenacious, we will be able to make it work."

Sarah adds, "But it's one hundred percent dependent on the retail price point, right? If you're a gardener you've probably noticed in seed catalogs that a pound of something will go for like $60 wholesale while a gram will retail for $3.50. Right? So that's like a thousand seed packets… okay, not a thousand but there's something like 450 grams per pound, so do the math. If you can sell retail by the gram then you're making significantly more money."

"Or at least you're not losing as much money," adds Andrew.

Sarah continues, "When we first got into the seed business, we were really disgusted by that math. And now that we better understand what the overhead and the loss and the risk is, we're not feeling any better about it. It's like a farmer getting five cents for a loaf of bread and the bread costs four bucks. That's what the wholesale seed world is like."

Andrew feels that it may be even worse than that. "Fundamentally, like many things in our society, it's about money and who controls it. It's like comparing mass-produced commodities to artisanal products that you know where they came from and that they're actually high quality."

Andrew emphasized that he and Sarah believe in producing seeds that people can save and reuse. But in the mainstream seed industry, most of the quality has been pushed off into hybrids, because they offer proprietary control. Individual companies control those seeds so farmers and gardeners can't save them. At the same time, the open pollinated seeds which people can save have been purchased by large corporations that are allowing them to degrade into commodity status so they won't compete quality-wise with their money-making hybrids.

The end result of these practices is that many of the traditional, open pollinated seeds are such poor quality now that they won't flourish in the field and they lack the nutritional profile they once offered. The Adaptive Seeds model is designed to reverse this trend.

"Our goal is to try to bring the value back to open pollinated and heritage seeds," explained Andrew. "To steward them up to their potential and make them thrive and produce better. Because it's kind of sad where the seed industry has gone, and it's time to try to put it right again."

Another key focus of Adaptive Seeds is their emphasis on the regional adaptability of their seeds. In fact, that's where their name comes from. When seeds are planted and plants are grown in a specific type of environment, the plants that do the best job of adapting to that environment produce the best vegetables. Naturally, the seeds those plants produce will have the best chance of succeeding when they are planted in a similar environment.

In order to carry out their goals, Andrew and Sarah understand that they've first got to make the business work, which includes making a profit.

Sarah, as ever, takes a practical view of that issue… "I've been thinking about how many more thousand dollars we need to make this year over last year. And asking myself how many hours of filling little seed packets does that translate to."

Many of the vegetable varieties Andrew Still and Sarah Kleeger grow are varieties that are popular in Europe but were rare or unavailable in the U.S. A lot of these seeds were acquired by Andrew and Sarah during their European travels as Seed Ambassadors.

Andrew, the entrepreneur, explains… "We have an operation that probably shouldn't be growing this fast, because it's really hard to finance and prepare for. And we're trying to self-finance it. It would be great to be in a position to not have to worry about cancerous growth every year, but we have to get to a critical mass as soon as possible to be a sustainable operation. We are close."

Then Andrew, the former philosophy major, expounds, "You know, this discussion is getting into the nitty gritty details of the business, which is kind of weird, but it's definitely part of it. It's an undervalued part of agriculture. I think a lot of people will get into farming thinking it's for the lifestyle, when the lifestyle is what falls out in the end. You have to have certain conditions in order for you to have that lifestyle. And having a business that works and knowing how to produce the food and knowing about your efficiencies and knowing how to market it and all that stuff is how you get to the lifestyle… but people think that you just have a lifestyle. Of course, all the things that get you to the lifestyle are actually part of what the lifestyle really is."

And Sarah concludes… "Now I would like to go for a walk."

And that's all part of being adaptive. Being able to adapt to whatever life and the market and any particular pursuit throws at you. Just like Sarah and Andrew. 🍴

(Clockwise from top left) Sarah showing recently harvested cabbage seeds. Andrew making some plant selection decisions in a row of coreopsis. One of the many seed production plots at Open Oak Farm.

ABOUT

ADAPTIVE SEEDS
Sarah Kleeger & Andrew Still
Sweet Home, Oregon

Adaptive Seeds was established in 2009 by Sarah Kleeger and Andrew Still and is based at Open Oak Farm, a 30-acre spread near Sweet Home, Oregon. Andrew and Sarah steward and disseminate rare, diverse, and resilient seed varieties for ecologically-minded farmers, gardeners, and seed savers.

Most of their seed is adapted to the Pacific Northwest and other short season northern climates. They sell only public domain, open pollinated (OP) seed, as well as many diverse gene pool mixes. All of their seed is grown by them on their farm or by a few regional friends who help with isolation requirements.

None of their seeds are proprietary hybrids (F1), patented, PVP, or genetically modified (GMO). All of their seed is grown without chemical fertilizers, herbicides, or pesticides. Adaptive Seeds was certified organic by Oregon Tilth in early summer 2013.

For more information about Andrew, Sarah and Adaptive Seeds, please visit their website at adaptiveseeds.com.

Deck Family Farm

*In addition to raising livestock in a healthy and humane way,
a major goal of John and Christine Deck and their children is to
help reestablish family farming as a viable economic endeavor.*

As I turn off High Pass Road to the Deck Family Farm there's a gate across the driveway. Cattle nose through fencing on either side, perhaps wondering what to make of this latest visitor. Unfastening the chain reminds me of the many pasture gates I opened growing up, and when I close it, like when I was a kid I try to chain it in a way that makes it easy to open next time.

The place looks like a working farm. A cluster of buildings on a bit of high ground. A mix of livestock braving the drizzling rain. Occasional water-filled potholes in the long, winding driveway. A dog standing beneath a shade tree barking at my car's approach. The line-up of work boots sitting on the porch beside the front door.

Based on the research I'd done, I wasn't surprised by anything I saw. But curiously, what did surprise me was the compelling nature of the conversation I encountered inside the house. Not that I was expecting it to be otherwise, but John and Christine Deck are genuinely intelligent people who, in a very comfortable and engaging way, shared some profound insights into our country's broken food system, why they believe a new farm movement is underway, and the part they're trying to play in all of that.

"I think people realize that the train we're on is going to crash," said Christine. "We live in an economy with externalized environmental and humanitarian costs. Finally, that discussion has moved outside the realm of economists and more and more people are understanding that we have to internalize those costs because we don't live in a world with unlimited resources. We can no longer base our economy on the belief that there's always going to be more... that there's always going to be growth. Sustainability depends more on not growing.

"We clearly need to develop some new paradigms, but they won't come easily. I honestly don't see any significant shifts occurring across society until there's some kind of collapse. Our government currently is a democracy of corporations. It's the time we live in, and money is dictating policy. But change has to start somewhere, and I think the organic and sustainable agriculture movement to some extent is about beginning that process. I feel like I'm doing my part. I'm trying my best to create a system I feel good living in and I feel can be sustainable. So I'll keep doing that."

Christine is clear about her intentions, but I wonder aloud if there is anything that would stop her from pursuing a farming life.

"Financial ruin. I couldn't do this without John, and he wouldn't do it without me. So we lean on each other pretty heavily right now, but finances represent our biggest challenge. We came here with no real debt, but after our first several years getting this operation up and running we had created significant debt, and we're working to pay that off. It's not easy, and I hope we get there. Not getting there represents the chief reason why we may not continue. But I believe we will get there, because one of our main goals is to help reestablish farming as a viable economic endeavor."

Not continuing to farm is probably one of the least

Winter bedding straw is stacked beneath a protective roof to the west of the Deck farm house. Bedding for livestock is important because most livestock, especially cattle, is protected from the Pacific Northwest rains. Keeping the cattle off the saturated pastures also prevents compaction of the soil.

favorable outcomes Christine can imagine. She is the first person in her family to graduate from college, and her degree from the University of California-Berkeley means a lot to her. But in her words, she comes from farm people, and it always has been her intention to farm if she could. That's not necessarily so for John, who admits that idealism has a lot to do with why he's now trying to create a sustainable farm. Idealism, family history, and Christine.

John and Christine first met at the beef barn and feedlot at UC-Davis. Christine was an animal science major at the time while John studied biology. And though John's upbringing was more urban, he was but one generation removed from a farming life and grew up hearing old farm stories from his uncles. As time passed, John and Christine developed a shared passion for farming and committed to giving it a go.

To help make things work at their current location, John brings his technical orientation to bear on the farm decision making. He still works for UC-Berkeley as a software developer… apparently a pretty good one because the university flies him all over the world working on special projects. So when he begins planning for the farm, he takes a scientific approach.

"I wouldn't call it science because I'd have control variables if it was scientific," said John, "but I am trying to be mechanistic about it, or at the very least organized."

After a fairly lengthy discussion of farm management that explored questions like how many pounds per square inch of pressure cow hooves exert on wet winter soil, the savings one can obtain by certifying farm land to grow livestock feed then allowing the animals to self harvest it, and the variables involved with getting an accurate measurement of the farm's overall dry matter production, John brings the conversation around to two things that he feels have a significant influence on making a farm sustainable.

What you can sell at market, and your ability to control costs… especially labor.

Market demand, he says, is what helped lead to their decision to raise a variety of different types of animals… demand and the need to break parasite life cycles. "If we're only selling beef then we kind of limit ourselves in terms of what we can sell, so we've learned to show up with a variety of products. Eggs are a good example. We decided to increase our egg production because we found a really strong demand for eggs. They help bring people into our market booth. So growing eggs is a really nice complement to our other products, and expansion of the egg operation kind of helped us ramp up."

John explains that in responding to the market demand for eggs, they increased the number of chickens that in the summertime are grazed on a set of organically certified fields of good quality grass with multiple water sources. Then the chickens are pulled off and sheep graze those fields during winter and early spring. This rotation is better for the fields, plus, it helps keep parasites from building up, which happens when you keep only one type of animal on the plot.

So when an adjustment is made in response to what will sell at the market, that change doesn't just show up in the market booth. An increase in the amount you're selling means more animals are needed on the farm, which affects the amount of resources (like land) that are needed. Which affects the operational system that includes all the animals, all the planting, the crop and animal rotations, the feed budget, and everything else. These aren't John's words, they're mine. But I think it's what John was trying to help me understand. There's a ripple effect that runs all the way through your operation and has to be accounted for.

I think this is why so many farmers have told me that one key to farming success is figuring out both what you can produce and what you can sell. One relies upon and informs the other. How you respond to either question has a significant effect on the farm's financial stability. Naturally, so does the cost of labor.

John points out that the family model can have a significant impact on this expense. "How many people you have living in the house and working on the farm makes a significant difference," said John. "In our society's prevailing paradigm, we're all encouraged to leave home. Everyone goes out and finds their own place to live, gets their own washer and dryer, becomes an independent consumer. But if people would choose to not try to develop individually focused investment paths, and instead work to build equity in a particular farm, the financial model changes and provides a significant boost to that small family farm.

"We have two kids living in town, paying rent, and attending school. What happens if they come back to the farm, live and work here, and their work is at least partially paid by the equity investment of the farm itself? That would really help out our equation on multiple levels, but especially in the fact that we're not having to hire other folks and pay workmen's comp, payroll taxes, insurance and all the other stuff you don't have to pay your kids. That may seem sad or a bit unfair to your children, but if it's treated as an equity investment, then that helps make it a positive for everyone."

Regardless of whether it's family or employees, Christine believes making a farm successful requires the commitment of a number of people.

"Right now, with our kids going off to college, we're relying heavily on interns," said Christine. "They are making it possible for us to keep moving forward. And although having a continuous stream of interns come through doesn't feel like a sustainable long-term solution, I am starting to believe that even if you don't have to look outside the family for all your farm labor, a healthy farm system often will. Because a pretty significant number of different types of people need to be involved to make things work well.

"It's funny, but the other day I was looking online for a

dunk tub for our sheep, and I found all these images – I think from the 50s – that had like four men putting a flock of sheep through a dunk tub. And there were other images with five guys, grown men, working around a cattle shoot. Things used to be different. It used to be that as an adult you would engage in this work at a serious level. And I think about now when I'm out there doing that on my own with my eleven year old daughter. I would like to see this farm support a lot of people, but to do that, everyone has to be able to give some things up… or trade one way of life for another."

Both John and Christine believe the trade offs are worthwhile. And they want to help as many people as possible gain the insights and understanding that comes from farm labor.

"I think the intern programs offer some really amazing opportunities," said John. "I see a lot of folks come through our farm who are spending several years working on three or four different farms. They get to see the seasons and really get a sense for it, and I really think it's an amazing educational opportunity. They may come away saying 'I don't ever want to farm again' but at least they have an informed opinion about where food comes from.

"We get a lot of customers who have no idea about what it takes to make food… all of the different factors that go into it. Many people look at labels and go 'oh, you're free range'

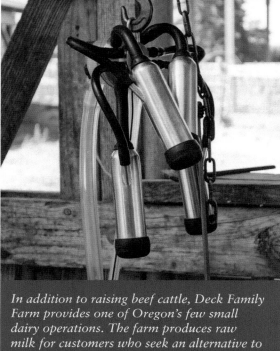

In addition to raising beef cattle, Deck Family Farm provides one of Oregon's few small dairy operations. The farm produces raw milk for customers who seek an alternative to conventional pasteurized and homogenized milk products.

or you're this or that and you can tell that these people don't have a clue about what those systems are really like. And regarding labels, if you're at the point where you have to distill food products down to labels, that's pretty bad. People need to understand it at a much more fundamental level. I think if you lived and worked on farms for a few years you'd be way beyond looking at labels. So for a lot of reasons, I'd certainly recommend that young people become farmers. I think it's actually revolutionary in some respects because it's outside the dominant paradigm. It's outside corporate America, and you're going to be running in the face of so many things in our society that you're certainly going to feel like a revolutionary."

Christine echoes John's passion… "It's highly satisfying to get into farming. It's hard work, but it means something. It's being out on the land. Improving it. And improving the lives of the animals we're caring for. I think bringing a product to market that is healthy is important. And is part of creating and maintaining a healthy ecosystem. I just feel like it's a right livelihood. It feels like I'm doing something that… well, it just feels good inside and it feels like it's making a positive impact on the world. And I guess I feel like I'm doing right by my people, too, by continuing a farming tradition." 🍴

(Clockwise from top left) Feeder pigs "cultivating" a grassy field. Lambs enjoying summer pasture. Office freezers filled with assorted meat products.

DECK FAMILY FARM
John & Christine Deck
Junction City, Oregon

The Deck Family Farm was founded near Junction City, Oregon in 2003 by California natives, John and Christine Deck and their five children. They offer pasture-raised meats at a variety of farmers markets, co-ops, and small grocery stores, as well as offering full and partial CSAs through their website.

The Deck's specialize in pasture-raised meats, including Oregon Tilth certified organic beef, which is 100 percent grass fed and finished; Oregon Tilth certified organic pasture-raised roaster chickens and eggs; pasture-raised and nut finished heritage pork; and spring lambs grown on dedicated pastures.

The overall objective of the Deck family is to create a life farming that sustains their family, improves the land, and supports the shift to a more sustainable approach to agriculture and family life.

For more information about the Deck Family Farm, please visit their website at deckfamilyfarm.com.

Your Backyard Farmer

Donna Smith and Robyn Streeter are living proof that an abundant amount of organic produce can be grown in urban and suburban yards of all sizes. All you need is a Backyard Farmer.

"We showed up at a client's home, and they had pulled every one of their pea plants out of the ground. So we asked them where the pea plants went, and they told us they were looking for peas. They thought they were in the roots, and we were like, holy cow, you don't know what peas are? So we told them they should go to the grocery store and look at the produce and figure out what all these vegetables actually look like.

"People don't know these things, you know. They know what food looks like when you eat it… usually. They go to a restaurant and maybe they get this great Swiss chard dish, and they know what that looks like, but they don't know what Swiss chard looks like in the dirt. So we realized that we needed to teach some more on that."

Robyn Streeter isn't making fun of her clients when she tells these stories, although she does laugh a lot while she's doing the telling. It's just her way of commenting on the fact that people today are disconnected from their food. Her business partner, Donna Smith, the more talkative half of this agricultural team, has even more stories.

"Sometimes when I've worked with kids I'll ask them where carrots come from," says Donna. "And they look at me like I'm crazy and say… 'from the store.' And I tell them, 'no, carrots don't come from the store, they come from somewhere else and then someone takes them to the store.'

"And then you take those kids through the process of growing carrots and when they finally pull one up, they're so excited because that's their own personal carrot and they clean the dirt off and take a bite. Then they understand where carrots come from."

Donna is clearly compassionate and understanding as she conveys that people simply don't know much about their food. For most people, food comes bundled with a rubber band around it. It's clean, and there are no bugs and no holes in the leaves. And that's the small amount of real food that people buy… you know, the stuff that's not in a box.

One of the goals for Robyn and Donna is to stay with their clients long enough to show them real food. Teach them from the beginning of the growing season all the way through. Even though it may take more than a single year, by the end, their customers truly understand and can even teach their neighbors how to put a tomato into the ground. And therein lies one of the many underlying benefits that Donna and Robyn bring to urban agriculture. They're not only growing food… they're also spreading information as much as they can.

Donna and Robyn met while attending the horticulture program at Clackamas Community College. Robyn recalls, "We met the first day of school and became friends pretty quickly. As we got closer to graduation, we liked the idea of doing something together, but we couldn't settle on what that was going to be. Eventually we decided to start a CSA so we started looking for property, but we kept running into problems."

When Robyn graduated a semester early, they still had

In addition to CSAs in the yards of clients, Donna and Robyn run a CSA operation in their own backyard.

found no land to farm, so she returned to her home in Idaho until something materialized. It was then Donna had her epiphany.

"It was one of those moments in life when you say, hey, why are we beating our head against the wall looking for land when everywhere we look, there's land all around us. So we just came up with a variation on Community Supported Agriculture (CSA). Usually, a farm produces and delivers a box of food. We eliminated the box. We just took ourselves right to the customer… to that family or group of families who own the food they hire us to put in the ground and harvest for them."

Donna started putting up flyers and immediately began receiving inquiries, so she called Robyn, who remembers it like this… "She called me and said 'hey are you in, because I'm getting phone calls from this.' So I said, I'm in."

Donna continues… "Two days later she's back from Idaho and we've already sold our first farm. The Oregonian got hold of that and wrote a two-page spread on us, which was really beautiful, and the day it came out we got over a hundred emails. Within six weeks we had 25 farms."

So these two urban farmers hit the ground running in 2006 when they launched Your Backyard Farmer, and they've farmed at least 25 yards every year since.

According to Donna, they learned a lot during that first year about a whole lot more than just how to grow food. Turns out that urban farming was so new, the State of Oregon wasn't sure there was such a thing. Running a business that had no legal definition turned into a challenge.

"We fought with the state about a lot of things, but in the end we managed to define what a farmer is and isn't. That definition didn't exist prior to us working through this process. Our problem was that even though we fell within federal guidelines, Oregon had no definition of what constituted a farm or a farmer."

In essence, by struggling through that process, Donna and

Robyn were able to legally protect themselves, and all other urban farming businesses, moving forward. Their pioneering effort did not go unnoticed, either here or abroad. As early as their second year in business they began receiving inquiries from people all over the world.

Robyn says, "We had people come visit us from Australia and from Spain. They actually came to Portland to hang out with us here, and we taught them what we do so they could take it back and do it wherever they were from. We've also helped people in the U.S., from California, Washington, other places. I think a lot of them are still at it."

Donna continues, "Seattle Urban Farms is doing it. Farmscapes in L.A. is doing it. Most of the people we've taught are still active. Green City Growers in Boston is doing very well. Then when we started getting calls from Barcelona, Spain and Hobart, Australia and these other places and we actually had people flying in and Robyn and I were like…"

Robyn: "What's going on?"

Donna: "This is really bizarre!"

Robyn: "And remember the Spanish magazine they put us in?"

Donna: "No, it was Italian."

Robyn: "Was it Italian?"

Donna: "It was Italian."

Robyn: "Italian magazine… it was like their main scientific agriculture organization and they flew two guys in to interview us and take pictures."

Donna: "And we're in the UK's urban planning guide about how you can bring your food right into your cities. It was bizarre to be propelled into the spotlight when we're both people who like to stay in the background, but we definitely have had a lot of fun with it. We still teach people, but we teach them in the off-season now, and a lot of the overseas people have seasons opposite what we do anyway. So we teach people to do this in the off-season rather than during our craziness."

Robyn: "Now, most of our time is spent farming rather than teaching."

Donna: "And we like spending time with our hands in the soil better than anything else anyway."

Exchanges like that are not uncommon when talking with this pair. Robyn's natural reticence momentarily fades and a bit of jocular chatter rises to meet Donna word for word. But soon her quiet smile returns and she sits back in her chair, content to simply take it all in.

Donna remains sitting forward, however, always willing to lead the conversation. In this case, she speculates on why so few urban farmers are able to duplicate and maintain what she and Robyn are doing.

"There have been a variety of people come into the Portland market and try to do what we do – we actually taught a couple of them – but they were in and out. This is very different from regular farming where you have the same piece of land to farm for 20 years and you take care of that soil and have something that's stable.

"In urban agriculture the way we do it, although we do have a number of farms we've been farming since 2006, we also have new customers coming every year, which means we are continually faced with a certain amount of unstable soil. I think that's a big difference."

Robyn expands, "It also takes a lot of organization. You've got 25 locations and each one of them needs something different. You have to remember each task, as well as keeping everyone on schedule so all farms are growing crops optimally."

Donna adds, "And you've got to consider microclimates. On a big farm there will be several places that have different microclimates that must be dealt with. For us, every single place we farm has different microclimates and soil structure. So urban agriculture has some unique challenges a lot of farmers don't want to deal with. Plus, they don't want to bring soil in to every new place each year. They don't want to carry their tools with them everywhere they go. In some ways this is like being a small contractor who hauls their workshop with them. There are a lot of unique requirements. And even we have our limits. If we had to do all year long what we do in the Fall and the Spring, we wouldn't be doing this, because that's not the fun part."

Robyn: "Soggy days aren't that fun."

Donna: "When you're hauling huge amounts of soil."

Robyn: "Moving soil eight hours every day."

Donna: "That's where it can be really physical work. We don't use any really big machinery so everything is done by hand."

Robyn: "Grunt work."
Donna: "And people go, 'wow, how do you do that?'"
Robyn: "You just do it."
Donna: "And then you're done with it. A lot of people don't want to work that hard. But I can't imagine, and neither can Robyn, doing anything different than what we do. Yeah, it's hard sometimes and we don't like each other sometimes, but ninety-nine percent of the time we do, and that's what it takes."

Your Backyard Farmer does extensive soil prep that turns typical yards into highly productive ones.

So these two women, year after year, keep putting in the work. The benefits are substantial. One fixed fee covers everything for a 37-week CSA running from the first part of March to the last of October. The farm agreement includes preparing the soil, setting up the trellising and water systems, the weeding, transplanting, seeding, and harvesting, as well as helping customers set up their compost systems. All this at a cost which makes it clear Donna and Robyn aren't running a get-rich-quick scheme.

Plus, because Your Backyard Farmer's customers keep all the food produced in their yards and no mechanized harvest or distribution energy is involved, the company's carbon footprint is actually quite small.

The things that Your Backyard Farmer don't do are any type of non-food-producing landscaping or raising animals. In both cases, the business would require different licensing. They can, however, offer suggestions.

Donna explains, "We don't do animals but we guide people on becoming more sustainable on their own property. That's our goal. We're going to provide you with your food source. In addition to that, we'll tell you, 'This is how you compost. This is how you raise chickens.' And because they don't have to worry about producing their food, we're giving them the opportunity to create some of these other things on their own. We can't do it for them, but we'll talk to them about it and help them understand what they need to do."

With all the health and ecological benefits available to people who own a farmable yard, it seems like the world could use a lot more backyard farmers. Donna and Robyn have pointed out some of the challenges, and they also emphasize that this is not an endeavor designed to make you rich.

Rather, it is a passion project. A way of life focused more on the journey than the gold. A perfect fit for those who love to get their hands dirty and work with nature to bring the world to life.

So for those who think this sounds like a love affair they could embrace, how could they get started? What steps should they take? Is it possible to become a backyard farmer anywhere in America?

Donna cuts right to the chase, "People should just call us. If they tell us they want to do this in their community, we're going to do what we can to help. There are a lot of ways to go about it. All of the people we've taught do things a bit differently."

Robyn says, "None of them are identical to what we do, which is how it should be. It's not a matter of this is how it should be, because every place has unique requirements."

Donna continues, "Portland's pretty progressive, so what we do may not work in more conservative communities. It can still be done there, but maybe a bit more contained. We can usually help people through the beginning stages, although they'll have to know what their community will allow. We feel like it's a fairly simple concept, but that's partly because we know it so well.

"It's time to stop thinking of a farmer as only someone who has 400 acres and a cow or 4,000 acres of corn. There are all kinds of farmers. Some of them are urban farmers, and the world could use a few more of them." 🍴

(Clockwise from top left) An ornamental garden plot that produces food. A large and lush suburban farm plot. Robyn (left) and Donna filling CSA boxes.

ABOUT

YOUR BACKYARD FARMER
Donna Smith & Robyn Streeter
Milwaukie, Oregon

Your Backyard Farmer is an urban farming business based in Milwaukie, Oregon, a suburban community located on the Willamette River and adjacent to Portland's southeast quadrant. Donna Smith and Robyn Streeter founded the business in 2006 and provide services to customers primarily located in the Portland, Milwaukie, and Gresham communities.

Your Backyard Farmer follows a Community Supported Agriculture (CSA) model in that it contracts with home owners to farm their yards for a single season for a fixed fee. In return, the home owner receives all the food that is produced. Customers typically have the option to renew the contract the following season. Your Backyard Farmer also offers a training program option in which they will train the home owners to grow their own food.

You can learn more about Donna, Robyn, and Your Backyard Farmer on their website: yourbackyardfarmer.com

Whistling Duck Farm

Located in the heart of southern Oregon's Applegate Valley, Whistling Duck Farm has built a legacy of organic farming innovation in this rugged and independent part of the the state.

Mary and Vince Alionis have been working together at digging in the dirt since the day they first met, which happened when they were working on a community garden project for the Green Party down in Dallas, Texas. I don't know what happened with that community garden, but the relationship Vince and Mary shared blossomed, and they began to look for a place to begin putting down roots.

After Dallas, they moved to California to help intensive gardening guru John Jeavons with a building project. Unfortunately, the funding for that project fell apart, so they began to look around for other opportunities. According to Vince, they met a lot of interesting people in California, but it just didn't feel like the right place for them, so they looked north and liked what they found in Oregon. More specifically, they liked the remote and rugged feel of southern Oregon along the Rogue River. So in 1991 they made the move to a nine acre plot in Shady Cove, Oregon, a doorway to Crater Lake country.

"We landed on nine acres of an old walnut orchard with a farm house and two wells, and that's where we got started," said Vince. "We immediately got a rototiller and worked up a little three-quarter acre spot and started doing any growers markets we could find. We gave ourselves two years to figure out what we were going to do longer term."

Two years later they were living and farming on forty acres farther up into the mountains on Elk Creek. "Cold country" is what Mary called it. They loved the ten years they spent there building their farm business. "That was a beautiful place," continued Vince, "a creek, a spring… a lions and bears kind of place, you know. Not necessarily a good production space, but an awesome homestead space."

There were challenges, though, like the shorter growing season found at higher altitudes. And the distance and difficulty of getting to markets. But it was the combination of a forest fire and the birth of their daughter that prompted these hardy farmers to seek a safer and more productive location somewhere in the valley. Which brought them to their current location on twenty-two acres of prime farmland beside Highway 238 in the Applegate Valley, where they've lived, farmed, and raised their daughter, Zosha, and son, Kazi.

Farming on highly productive soil in the valley was quite different from farming on marginal soil on the upper Rogue. So when Mary and Vince continued to apply the same methodologies they had developed to accommodate previous challenges, they were faced with more produce than they knew what to do with because everything grew so well. The abundance enabled them to quickly expand their markets, as well as helping to pioneer better farming techniques in their part of the state.

"Coming from the colder climate, we had been forced to develop techniques that simply weren't being applied down here in the valley," explained Mary. "There was no one doing greenhouse culture or intensive succession plantings. And only a couple of farms were beginning to stretch the season. These

The Siskiyou Mountains rise behind Whistling Duck Farm greenhouses. Vince and Mary Alionis were among the first organic farmers in this part of the valley to begin using greenhouses to extend the growing season for many vegetable varieties.

were all well established methodologies, but it was just that people weren't doing it here because they didn't have to. Plus I suppose it's like anything else… people don't do it until they see it happen, and being on the highway, we're very visible. So we feel like we were able to bring some positive influences to growers down here."

Vince laughed as he continued the story… "Of course we flipped out all the locals when we showed up because we immediately started putting up greenhouses and preparing the soil and selling product. When we landed here, there was no house. We were living in a yurt up in Williams. But we got here in December and had a crop out of the ground by April."

"But we understand that the real pioneers were the people

up in the Willamette Valley doing things like this in the mid to late 80s," Mary acknowledged. "We actually gleaned some of our information from them, and of course anyplace else we could find it. It wasn't easy to get information back then. There wasn't any internet."

Both Vince and Mary believe that successful farming requires compulsive entrepreneurship. The way they developed Whistling Duck shows that. They also believe that nobody survives as an entrepreneur unless they're willing to do whatever it takes to get the job done, and that you can tell pretty quickly who's going to make it and who's not based on their willingness to do that. But as I listen to them tell their story, it sounds like information and flexibility might be just as important as hard work. The ability to get the information, figure out how to apply it to your own operation, and adapt your plans and processes based on what you learn and what you encounter.

They both like to use the term "game changer" when they talk about new discoveries or new ideas. Like how their discovery of the Allis Chalmers G cultivating tractor was a game changer that enabled them to farm more efficiently on larger acreage. Their refrigerated truck was a game changer because it served as a walk-in cooler when nobody around them had walk-in coolers. "That truck gave us a real qualitative edge," said Vince. "We were always keen about getting product out early, getting it in water to get the field heat out, and then getting it cold. That extends shelf life, which means our customers were getting a better quality product."

And though they didn't use the term, Vince being diagnosed with cancer several years after their move to the valley also was a game changer. It meant that he would be doing less, Mary would be doing more, and their need for good quality, reliable workers would increase. It also meant that their long-term plans for the farm would need to be adjusted to remove some of the stress.

They've been lucky with their crew. "We have about ten people right now, and three or four are year-round employees," said Mary. "That doesn't mean they're full-time during the winter months... more like twenty to thirty hours on average, but they are year-round and they've been here for quite a few years now. If we lost our key people, I'm not sure what we'd do. But for now, we're good."

To help expand their year-round workload and keep workers busy, Whistling Duck added seed garlic to its product list. They grow approximately twenty varieties of garlic, which requires monitoring and care throughout the winter. Vince commented that seed garlic is their export crop, meaning that the majority of it leaves their valley and new money comes into their valley, which is critically important to the local economy. In fact, organic seed production is becoming a key economic driver in the Applegate and Rogue Valleys, especially since both Josephine and Jackson counties voted to ban GMO crops, which can destroy organic seed operations.

Another year-round endeavor is their farm store. They currently maintain a self-serve farm store in a small building beside the highway, but a new and much larger one is being constructed inside an adjacent barn. Plus, they've recently launched a new fermented foods business.

"Now we make kraut," said Mary, "so that's another winter gig with a lot of winter cropping. It's also an important part of our long-term plan. We took over a kraut business about two years ago from a friend named Kirsten Shockey. She wanted to get out of the business of making and retailing fermented foods and concentrate on writing and teaching about fermentation. And her new book, *Fermented Vegetables*, is great by the way. And this has turned out very well for us because it doesn't get any easier to start a kraut business than to have an expert give you theirs. How can you turn that down?"

Taking on a new operation this time intensive did mean that Mary put on one more hat when she already was probably wearing too many hats. But both Mary and Vince are very upbeat about the potential this undertaking represents. And it definitely improves their conversion ratio.

"A couple months ago I had several hundred pounds of radishes that all came out at once because someone had seeded too many," said Mary. "But I realized I could ferment them, so we made about twenty-five or thirty gallons of fermented radishes of various kinds. And they're incredible... just so good. And that's something we wouldn't have had a market for and would have tilled in. Now we're turning it into value added products that will sell."

Vince commented that the fermentation path also fit well with their existing approach from a nutritional standpoint. Whistling Duck has been catering to the juicing and smoothie crowd for a number of years by growing nutrient dense veggies and greens like lambsquarters and purslane. So fermented foods fits right in. "It's a great market," he said, "very cutting edge. We're doing something creative in the kitchen, but we're not encouraging people to buy jams or cookies. It's fermented. Probiotic. It's good for you, so there's nothing negative about it."

Mary says they have no problem selling their krauts and fermented products. Because they had been selling the products before they took over production, they had a built in customer

Fermented foods are rising in popularity because of their nutritional benefits. Whistling Duck is well positioned to benefit from this trend after assuming control of a local fermentation business and creating a value-added outlet for much of their produce. The next step is adding a certified kitchen to their new farm store to handle their ferments and other value-added endeavors.

base, which has been steadily expanding along with the rapidly growing fermented food market nationwide. Mary also has found that creative marketing can bring significant benefits.

"We're creating so many different ferments and coming up with our own recipes," she explained. "One of them I put together recently was an attempt to make something seriously good for you. It had root parsley, burdock, nettles, turmeric… a lot of medicinal stuff. Unfortunately it didn't taste very good. And I couldn't just throw it out because there was nothing wrong with it. Instead, I cut it in with some plain kraut to dilute it, called it 'the healer' and charged more for it. It sold out immediately."

In addition to it's market potential, both Mary and Vince are excited about the fermentation business because of the role it plays in their new five year vision.

"Our end play is to get our store developed so we can scale back the super intensive farming gig," said Mary. "We want to have a local store that sells our products and products from all sorts of other local people. Our five year plan is to not be doing all these markets and not be doing wholesale, and instead keeping it all right here. Keeping our seed garlic, and still growing veggies, but selling them here in the store. We have a tendency to just keep doing it all, but we need to scale back because we're getting old. And if we simplify things, that also will make it easier to find people who can keep it going."

Mary and Vince make no bones about how tough it

might be to find the right people to help run their farm and farm store. But like with everything else they've done since they got started back in 1991, they're trying to keep an open mind and look at all the possible alternatives.

"This property is actually two tax lots," explained Mary. "The house sits on one tax lot and the fields are another one. And I would love to get things set up so we could build a house on the field lot and have another farmer working here with us… have someone else living here on the farm and provide housing for them. Because housing is usually the sticky wicket. And I'm not talking about just a bunk house, but a real home so another farmer or farm family could live here and work with us. Who knows what that would end up looking like. I'm not into a business partnership, but I wouldn't have any problem saying, okay, we own this property and we will lease it to you, and you run this aspect of a farm. You do the green vegetable production or whatever. They can run their aspect of the farm, and we can run ours. It may not go that way, but I'm good with all that. The main thing would be that we're interdependent but we remain individuals."

In the meantime, Whistling Duck Farm will continue down the path it's been on since it got started. And Vince and Mary Alionis will keep digging in the dirt together, just as they were when they met so many years ago. 🍴

(Clockwise from top left) Interior view of Whistling Duck Farm's new farm store. The recently renovated barn that provides Mary and Vince with a high quality on-farm retail store. A farm building with radiant heated slab, solar alignment, and a 10.34 kw solar PV system that provides power to the full barn and propagation house complex. Photos courtesy of Mary Alionis.

ABOUT

WHISTLING DUCK FARM
Mary & Vince Alionis
Grants Pass, Oregon

Whistling Duck Farm was founded by farmers Mary and Vince Alionis in 1991 and moved to its current location in southern Oregon's Applegate Valley in 2003. The farm's twenty-two acres are certified organic by the Oregon Department of Agriculture.

Mary and Vince grow a wide range of produce for their farm store, local growers' markets, fine restaurants and natural foods stores. They also trial and grow certified organic gourmet seed garlic, both for local markets and for farmers and gardeners across the country via their website.

The latest adventure at Whistling Duck is their new fermented foods venture, which features both traditional and unusual ferments, all of which will be offered at their new on-farm retail store.

For more information about Whistling Duck Farm, please visit their website at whistlingduckfarm.com.

Peace Seedlings

Peace Seedlings' catalog organizes seeds according to their scientific relationships to one another, which is one of many unique characteristics of this remarkable source for organically grown seed.

Most people who closely follow the organic farm movement in this country have heard of Dr. Alan 'Mushroom' Kapuler, the plant-breeding pioneer who was one of the original founders of Seeds of Change and who helped put environmentally adaptive, public domain plant breeding on the map.

What fewer people may know is that for that past seven years Alan Kapuler has been working in his Corvallis, Oregon breeding garden with two dedicated proteges... his daughter, Dylana Kapuler, and her partner, Mario DiBenedetto.

Dylana and Mario have spent these years formally studying with, being inspired by, and working with a master plant breeder who, in turn, has trained, nurtured, and guided them through the equivalent of a degree in plant breeding and garden ecology.

Today, these two proteges are beginning to realize what they have acquired – the ability to help sustain Dr. Kapuler's work and, in turn, build a seed company of their own. Rather than take over the senior Kapuler's Peace Seeds, Dylana and Mario launched their business, Peace Seedlings, with an obvious nod to their mentor.

"Some people were surprised when we didn't just take over Peace Seeds," explains Mario, "because people knew Mushroom was at a place where he was thinking about retiring. But he didn't want to stop doing everything, he just didn't want to continue selecting all these varieties he had developed. So we started taking that on. As we did that, we also got more involved with our own breeding projects, and

naturally since we were working with Mushroom, our work was focused on public domain plant breeding."

"Which needs to happen for the sake of adaptation," adds Dylana. "The climate is changing, so seeds need to be selected based on where they're located and how they adapt to those changes, as well as a number of other considerations like nutritional makeup. My dad's past work needs to be continued and built on. By doing that, we're freeing him up to focus on his latest inspirations, which include things like native food plants that are overlooked or neglected. He's inspired by species that are hard to get your hands on... that you have to wild collect and learn how to grow so they're not at the mercy of who knows what bulldozer."

Mario continues, "We have two different seed companies, but obviously there's still a lot of collaboration, especially on varieties he developed. As we grow and develop plants, we talk with him about what he was selecting for, what his process was. If we weren't doing this, then things like double red sweet corn would no longer be available. And that would be kind of sad."

As I talk with Mario and Dylana, the influence of Alan is ever present. In his quiet, humble manner, Mario speaks with something approaching reverence about what Dylana's father has taught him, how he has inspired him to embrace the principles of breeding plants in a way that benefits all people and selecting for traits beyond the lure of profit.

"We visited our friend at Oregon State not long ago, and they were doing berry trials," Mario says. "They have these

Peace Seedlings' greenhouse is home to an indoor forest of citrus trees and other tropical plants.

amazing, huge raspberries, blueberries, honey berries... all sorts of berries. But the only things the researchers go by is how sweet, big, and productive they are. That's cool. That makes sense. But Mushroom's thing is that if everyone is stuck on sugar, what are we doing to ourselves? We might want to look at other aspects, like amino acids or anthocyanins. Things that might make the berries more beneficial. We should keep trying to learn what's possible.

"And honestly it's pretty hard not to be inspired to learn when you're around Mushroom," Mario explains. "One of the most interesting things about him is how incredibly motivated he is to learn at all times. So he's been a huge help for us, just to keep us inspired, and also to give us direction and things to learn. He continually challenges us. Plus, he advocates that if you want to talk about something then you'd better learn as much about it as you can, otherwise you're not really doing other people justice... or yourself for that matter."

I ask if it has always been this way for Dylana, growing up in her father's gardens and under his watchful eye.

"As a young kid I was certainly encouraged to pay attention to seeds and plants," says Dylana. "He always encouraged all of us – me and my sisters – to be interested in plants. For me, it probably really took hold when I didn't go to kindergarten because I wanted to stay home and watch my dad clean seeds and hang out, and that was fine with him. Of course, after that I did go to public school, which pulled me into that whole public school world for a period of time, but I still spent time in the garden. At one time I thought I wanted to be an organic farmer and work for Seeds of Change... of course that was before Seeds of Change was bought by a large corporation.

"And then there was a time as a teenager when I kind of forgot all this and spent some time asking teenager questions, like 'what am I going to do with my life' but by the time I was halfway through high school I was backyard gardening a lot on my own, and I knew I wasn't going to college. Because I just wanted to learn hands-on. I'm really a hands-on learner."

While Dylana was growing up in her Corvallis gardens, Mario got his start a little farther north in Washington. He grew up on the Olympic peninsula. Like Dylana, he doesn't have much interest in higher education, although he did have an Associates degree through a local community college by the time he was eighteen. But a short time later he relocated to Eugene, Oregon and began working on the Walama Restoration Project, a non-profit dedicated to environmental stewardship and biological diversity through education and habitat restoration. It was then he and Dylana met through mutual friends and first got together. And not long thereafter, he moved to Corvallis to live with Dylana and her parents.

Dylana puts it simply… "and that's what led to us being out here in the garden."

Mario adds, "And we began gardening and learning. That's what Mushroom has always advocated for, just learning about plants, and he feels one of the best ways to really learn is to get hands-on experience with as many different plants and seeds as you can. That grew as kind of an organic process. You're saving seeds, you're learning more each year, and then you reach a point of realizing that you have quite a bit of experience, a growing body of knowledge, and a whole lot of seed."

"When we started, my dad said it would take us five years before we really knew anything," says Dylana. "At the time I was like, five years? But he was right. You have to go through that many cycles to really get an understanding of what's happening and why. So I guess our first five years was kind of like going to college."

One very important thing both Dylana and Mario have gained from their informal apprenticeship is a broad-based foundation of practical field experience they most likely would not have garnered from a university degree. The agricultural departments of most land grant colleges have followed the money represented by industrial agriculture in the same way businesses do. This is because university research dollars come from the large agrochemical companies or the government agencies run by former agrochemical company employees. Because of this fact, they reflect the industry they are supporting, which is an industry of specialization.

When agriculture is run as a mechanized industry, it functions in a manner similar to a giant assembly line with every person doing their one specialized task. For that reason, a college student studying plant breeding might spend his or her entire time focused on a single factor within a specific market. Alan Kapuler recognized that and thus encouraged actual field work.

"To us, college just doesn't make much sense," says Dylana. "First, of course, there's the debt. Why would we want to go into debt to do something we can do right here? And from what I've seen, those kids are completely focused on just one type of something the whole time. One strain of barley. Or barley just for beer. Or barley just for something else. They spend their whole time at the university working on some professor's project that's not even their own work and they come away with this narrow bit of knowledge about a subject as broad as plant diversity. With virtually no field experience. Doesn't make sense to me. I can definitely see why my dad gives what we do such props."

Mario adds, "And the resources they waste. When you go look at their greenhouses, the lights are on in the middle of the day and they're growing oats in the wintertime. There's so much space and energy and resources being put into it, I hope it's doing someone some good, because otherwise it's just unforgivably wasteful."

"But we're mostly talking about plant breeding," explains Dylana. "About getting real experience living within a truly diverse ecosystem and working through each year's cycles and watching what really happens in this particular environment. Seeing which plants succeed and how they adapt. We do understand that universities offer some very important training that you can only get there. We talked earlier about analyzing amino acids or other nutritional qualities… we can't do that here and that work needs to be done even more than it is now."

Mario continued with Dylana's comments about studying what is happening in front of you… paying attention to how plants are adapting in nature. It was a point both of them kept coming back to… allowing plants to adapt to the environment. The breeder's goal, they believe, should be to limit external inputs, which include things like excess fertilizer, to the extent possible and allow plants to grow by themselves within a diverse ecosystem. Some plants will do better than other plants. So if the seed from successful plants is saved, the grower will have a better chance of succeeding with that plant the next year because the parent plant was already successful within those growing conditions. Every farmer wants – or should want – plants that are vigorous without high inputs because it makes good sense, both ethically and economically.

Another way Mario and Dylana are attempting to bring economic benefits to the marketplace are through the introduction of new types of food plants, especially root tubers from the Andean region of South America. They have been growing yacon, oca, and mashua for a number of years, and because the plants are so prolific, they have become Peace Seedlings' biggest product, both in sales and in weight.

"The Andean people had an amazing food culture," says Dylana. "They grew more root vegetables and different taxa than anybody. I mean potatoes came from them, of course. But they also grew oca, which are tuberous oxalis, mashua, a tuberous rooted nasturtium, and yacon, which is an edible rooted daisy related to sunflowers and such. And oh my gosh, the list just goes on."

Mario adds, "They have something like a dozen different tuberous rooted food plants in a dozen different families. I feel they are some of the best farmers and food plant developers in the world. Andean foods have provided important staple foods to cultures all over the world."

Adding these unique root vegetables to its product line

Cacti and succulents of all types were among the first plants to capture the imagination of seed breeder Dylana Kapuler. Years later, Peace Seedings continues to offer them.

provides Peace Seedlings with a revenue source that helps make their business more viable for the long term. Which is important, because now that Mario and Dylana have gained the ability to successfully maintain heirlooms and introduce useful new varieties of seed, they're still novices at the business of running a business. Currently, like their personalities and outlook, their business practices are unconventional, even with seemingly straightforward decisions like how many seeds to include in each packet they sell.

Mario explains, "Most seed companies state how many seeds are in each packet, and they include pretty much exactly that number. We list a quantity also, but we view that as a minimum. If we've got plenty of seed, we'll just add extra to each packet. I mean, you don't want to waste good seed. And usually, we've got plenty. We just list the minimum in case we don't get a good crop or a bird swoops in and eats several hundred starts… which happens sometimes. That's just kind of how we do business. I guess there's still a lot we need to learn about the business side of things."

After a short discussion about marketing topics, I ask them both if they plan to do this forever. The answer shouldn't have surprised me.

"We're not really the type of people to make that strong of a statement," said Mario. "In general we see ourselves doing this type of work, but we see everyday how much adaptation is just a part of this world, so we understand that applies to us, too. So it's hard to know. But we hope that we're able to keep this land and keep doing this work for a lot longer." 🍴

(Clockwise from top left) A colorful variety of zinnia. Circular rows add interest to Peace Seedlings' test gardens. Mashua plants are one of several South American tubers grown here.

ABOUT

PEACE SEEDLINGS
Dylana Kapuler & Mario DiBenedetto
Corvallis, Oregon

Peace Seedlings founders are second generation public domain plant breeders and seed saving stewards. Dylana and Mario continue the work and ever growing legacy of Peace Seeds and its founders, Alan and Linda Kapuler. Peace Seeds and Peace Seedlings are separate companies although they work out of the same gardens and collaborate on many crops, but each have unique offerings.

Peace Seedlings is committed to breeding and adapting new vegetables for our time and to looking beyond the food plants that dominate our current food system for greater diversity to sustain us. One example of this is their focus on growing and offering South American food plants and tubers.

Though not certified, all Peace Seedlings products are grown organically in a three acre plot located near Corvallis, Oregon. They provide many heirlooms and a growing number of original public domain offerings. All seed varieties are open-pollinated. They grow, save, clean, and pack all the seed they sell.

For more information about Peace Seedlings, please visit their website:
peaceseedlingsseeds.blogspot.com

Lonely Lane Farm

Lonely Lane Farm, which has called Mt. Angel, Oregon home for three generations, has reinvented itself as a natural, sustainable operation and added a meat processing capability to remain financially viable.

Mike and Patty Kloft and their one-year-old son, John, are a young family working hard to save their way of life and continue along the path of their forebears. Their home is a traditional family farm. The kind that's disappearing at a breakneck pace. And they are traditional farm people. The kind you might have met twenty or forty or even sixty years ago.

Growing up on a family farm in the 1950s afforded me many opportunities to meet just such people, and after spending a couple of hours with Mike and Patty, I realized I could just as easily have met them back then. Their farm is their life. It defines them and influences every thought they have. And one of the most important aspects of their approach to farming reflects their approach to family… how do they make what they've been entrusted with even better for the next generation? That's a question which lies at the very heart of sustainability. But it's how you answer it that really matters.

Mike and Patty grew up less than three miles apart in the countryside near Mt. Angel, Oregon. Mike's family started their farm in 1939. Patty's dates back to 1890… same family, same farm all that time. There's a lot of history packed into those years. But times change, and by the year 2000, as Mike was coming into his turn at running the farm, he was wondering if there was going to be any farm to run.

"Things had gotten to the point where we just weren't making enough money to sustain everyone anymore," said Mike. "My grandfather started out with a dairy, and then in 1985 they sold their dairy herd and just ran beef cattle. That worked well enough for awhile, but after about fifteen years, we knew it wasn't going to last. I was going to college down at Oregon State studying ag around then, and I was wondering if I was going to have to get out of farming."

Fortunately Mike signed up for a class about world foods and the cultural implications of international agriculture which was being taught by OSU Small Farms program director Garry Stephenson. I don't know how much Mike remembers about international agriculture, but he clearly recalls a conversation he had with his professor.

"After class one day Garry and I were sitting around talking, and I told him I wasn't sure what I should do about our farm because we weren't bringing in the amount of money we needed," said Mike. "I was out of ideas, but he asked me to tell him what we do. So I explained it and he was like, 'well, it sounds like you're raising everything sustainably.' And I had never even looked at it that way, but it was true. We were and always had. We just didn't know we should be marketing it that way."

As Mike lays it out, his family always had run a sustainable farm focused on producing healthy, quality products. For them, that meant controlling all of their inputs by producing them on the farm.

"We produced our own livestock, our own feed, our own bedding… everything was right here," Mike explained. "We used no hormones, no antibiotics, no animal byproducts. We'd

The countryside around the community of Mt. Angel has long been home to many farms similar to Lonely Lane. Patty grew up on a farm three miles away that dates back to the late nineteenth century. The very old oak tree in the center of the photo inspired the name of Mike and Patty's meat processing business, Century Oak Packing Company.

always raised our own GMO-free feeds. Garry pointed out that we were obviously passionate about it, or we wouldn't be doing it that way, and he was right. Turns out it was more a matter of us finding the market that fit our farm than it was trying to change our farm to chase some other market. I actively promote that approach to people now, and I know Garry does the same. So that's how we got started in the sustainable farming market, and by 2001 we were able to get into our first alternative grocery stores."

Over the following ten years Lonely Lane Farm continued raising sustainable beef, getting it processed at Mt. Angel Meat Company and selling to stores, a few restaurants, and direct to consumers at farmers markets. But that doesn't mean

there were no changes on the farm, and Mike would probably say the first big change was the most important. It all started innocently enough while preparing for a farmers market.

"Patty's dad was raising pigs for us," said Mike, "and I was talking to him one day and said we needed some extra help at the farmers market. So he said he'd check to see if any of his girls wanted to help. Patty's older sister wasn't interested, but Patty said okay, and that's actually how we got to know each other."

Patty explains that she's about twelve years younger than Mike so they'd never really had an opportunity to get to know each other until they started working together.

"We were talking about that last night," said Patty, "how

it was nice to work together and become good friends before we started dating. Then after a few years, we decided to get married."

"You stumble on something good," adds Mike, "and you're lucky you do."

I'm sure a part of the personal compatibility of Patty and Mike can be attributed to their shared values, because Patty comes from the same type of farm as Mike. The only real difference is the fact that Patty grew up on a hog farm, and it's something she likes to talk about. For the majority of our interview Patty let Mike do most of the talking until I asked her about pigs – what breeds they raised – and she perked up.

"They're actually a mixture," she said. "For a long time when I was growing up as well as raising fat hogs we also did a lot of 4-H. And it seemed to us that they were kind of into colors. So we actually bred for color. To do that we just changed our boar out every year and picked a different breed. So we have everything in there from Spotted Poland to York, we did Hampshire, Duroc. We always kept our own sows but over time we ended up with quite a mix."

I shared with her that growing up my family raised mostly a hampshire-yorkshire-duroc cross, but that a friend of mine raised Berkshires. And she responded in a way only a true farm girl could… "They are such pretty pigs. I always liked Berkshires. We had some of those around, too."

After our brief homage to pigs, we returned to the subject of Lonely Lane Farm and the other significant change that took place following the move to marketing natural, sustainable meats. The development of an on-farm meat processing plant. If that sounds like a pretty significant development, it should, because it is. But from Mike's perspective, it's all about vertical integration.

"We always had talked about being vertically integrated," said Mike, "which for us meant that if we had a cow herd we would be producing our own animals and all of our own feed. So we were doing cow-calf all the way through finished product with all our own inputs. I thought we were vertically integrated. But when we made the switch to this market, the product goes directly from our farm to the consumer. And I realized we weren't completely vertically integrated because we could raise a phenomenal product to a certain point and then we had to turn it over to someone else and see what we got back. And we did that for awhile."

Then Mike stumbled into another opportunity that opened the door to what turned out to be a long but fruitful journey. Without realizing what we was setting himself up to do, he managed to negotiate his own processing space in an existing meat processing facility by agreeing to assist them with some USDA planning requirements.

"That really helped us," explains Mike, "because we were actually processing our own meat in that facility one day a week. Learning how to do it, and learning everything else that went along with it. We went on that way for awhile, but before too long, we realized that we were going to need more control and the ability to expand our operation if we were going to make it work. Either we needed to fully commit to the processing side or get out of farming. For us, it needed to be an all or nothing deal. Simple as that. So that took us in the direction of looking at processing facilities."

The Klofts' search for a processing plant included both existing facilities that were closed, with the possibility of reopening, and those still operating, but with owners nearing retirement. But after a couple near misses, they finally settled on converting buildings on their farm and creating their own plant. From that point, the focus was on determining what all needed to happen to get from point A to point B.

"Fortunately we'd been working in this area for awhile and we had built a pretty good rapport with the USDA inspectors," said Mike. "I was able to call them and ask them what they were looking for in a facility. I mean, I can read the regulations, but I really wanted them to come out and walk through the space with me to see if what I was thinking would actually work."

Mike and Patty weren't overly reassured when the visiting inspector couldn't visualize a plant working the way Mike explained it. He admitted that the plan sounded right, but the best response he could come up with was to just try it and see how it goes. Undeterred, the couple recruited family members and began converting an old dairy barn into a meat packing house.

Century Oak Packing is located in a converted dairy barn where Mike's grandfather used to milk cows. Today, it serves as a processing center for beef cattle raised at Lonely Lane, as well as pork, lamb, and goat raised on nearby farms.

Now if you're wondering what the term family farm really means, here's a good example. The principal players in the construction of what would become Century Oak Packing were Mike, his father, his uncle, and Patty. Plus, there was a cousin who just happened to be an engineer. And long story short, they got it done. The USDA came in; the facility passed scrutiny; and the Klofts were assigned a USDA inspector. Now they run the packing house five days a week year-round, providing employment for a crew of local community members and a retail meat-cutting capability for surrounding farms.

Now that the meat processing is running smoothly, both Mike and Patty are happy about being able to spend more time farming again, because that's what they love the most. Both are open about the fact that they always wanted to continue life on the farm.

"For me personally, I always wanted to farm," said Patty. "I just didn't know exactly what form it would take to be honest with you. But after Mike and I started dating, I knew I could do this for the rest of my life. And now with our son, we want to make sure he gets to grow up in a similar way. He loves being outside, watching the animals, splashing in the creek. That's country life."

And though he doesn't say so explicitly, I have to believe Mike already holds out hope that someday he will turn the reins of his farm over to his son, John.

"I will say that's one thing about the farm we're on," Mike shared, "that beginning with my grandfather, they've always seen it as being a steward for the land for their own generation. Doing whatever they can do to improve the soil, the environment, the quality of life… everything they've got for the next generation that's coming along. My grandfather did that for my dad and was willing enough to turn the reins over and let him go the route he wanted to go when the time came. And my dad did the same thing for me.

"I think that's where we lucked out, because a lot of people aren't that fortunate with successive generation farms. You know, sometimes a generation will get stuck in what they do and think that's the only way to do it. They don't realize that situations change and markets change and that you have to adapt. They don't take the long-term view and try to make decisions that strengthen what they have rather than only trying to get bigger or just going for the money right now. So I've had a good example to learn from, and I feel like I'm really lucky that my dad told me, 'I picked what I did for my generation, now you pick what you do for your generation.'"

It's easy to believe that Patty and Mike are going to do whatever it takes to make sure young John has the same opportunities they've had. ✄

(Clockwise from top left) Cows are brought in to eat while their calves wait outside. Silage in an open bunker and various types of hay in a nearby shed are produced on the farm. Mike's father still offers a helping hand.

ABOUT

LONELY LANE FARM
Mike & Patty Kloft
Mt. Angel, Oregon

Located in Oregon's northern Willamette Valley, Lonely Lane Farm is a family owned and operated sustainable farming operation. Founded in 1939, their farm has remained devoted to family-oriented values as it's management was handed from one generation to the next.

Owners Mike and Patty Kloft pride themselves on providing sustainably produced fresh and healthy meats to customers through farmers markets, local stores, and restaurants. In addition to their livestock operations, Mike and Patty produce alfalfa and silage for feed, and wheat, which they sell on the commodity market and which produces the straw they need for livestock bedding and composting.

In addition to their farming, they also have developed and operate Century Oak Processing, an on-farm meat processing plant that serves their production needs as well as those of neighboring small sustainable farms.

For more information about Lonely Lane Farm or Century Oak Processing, please visit their websites at lonelylanefarms.com and centuryoakpacking.com.

OSU Center for Small Farms and Community Food Systems

Oregon State's Small Farms Program has been a primary catalyst for the growth of Oregon's new farm movement, both by training new farmers and supporting the efforts of existing farms and organizations.

Garry Stephenson wants to change the world. He claims he's wanted that forever, and though he kind of chuckles when he says it, I don't doubt it one bit. His is a sentiment shared by many who witnessed the original "back-to-the-land" movement of the late 60s and early 70s. Garry also shares another sentiment with many from that era... that time is getting short if any significant changes are to be realized in this lifetime. That's why the creation of the Center for Small Farms and Community Food Systems is so important to him. Garry directs the Center and is assisted by Associate Director Lauren Gwin.

"You know, what astonishes me is that in many ways, we're back to dealing with the same issues we dealt with in the 60s," said Garry. "But what I really like about the current young farmer movement is, unlike our movement where people just wanted to leave society and try something new, these people want to start businesses that will change the world. They want to make a living at this, and they want to do it in a way that's not going to contribute to the chemical industry's bottom line. So it's a twofer. And how the hell did they think of this, and we didn't?"

Regardless of how it happened, the commitment within Oregon's (and the nation's) new farm movement to strong local and regional economies, healthy food, and family farms is worth supporting. And that's where Garry and Lauren's new Center comes into play.

To help put the creation of this Center into perspective, it might help to understand what a land grant college is, and how the agricultural extension service works. Oregon State is a land grant university, which means it was originally created as a result of federal legislation designed to help states develop colleges that focus on agriculture, science and engineering. Basically, OSU started out as an 'ag' school.

One way land grant schools imparted their agricultural research and acquired knowledge to real world farmers was through an 'extension service.' Extension agents, who were employees of the university, were located throughout the state to help make them accessible to working farmers. And as new ideas and techniques were developed by university researchers, extension agents helped farmers implement those new ideas. It all was done in the name of progress.

As long as the ideas coming out of the universities are good ideas, this system works well. But because the universities rely on outside organizations for much of their research funding, the system is vulnerable to the influences and ideas of people with money and motives that aren't always altruistic.

Following World War II, the businesses that had a lot of money and the idea of creating a new type of chemically dependent, industrial agriculture were the chemical companies that had been in the business of making bombs and chemical weapons. With a greatly reduced need for weapons, the chemicals were repurposed to become chemical fertilizers and

Jen Aron (left), Learning Gardens Laboratory Manager, and Weston Miller, OSU Extension Faculty, are the lead instructors for the Beginning Urban Farmer Apprenticeship program. BUFA is a partnership between OSU Extension Service and Multnomah County designed to provide in-depth and comprehensive training in sustainable, small-scale, urban farming methods. The program exemplifies OSU's collaboration with other organizations striving to advance small farms.

agricultural poisons (pesticides and herbicides). Land grant universities then were essentially hired to do research on how best to use these newly available chemicals to enable farmers to grow more food and make more money. And of course, eventually this research was expanded to include gene research for genetically modified seeds, which weren't modified for health reasons, but rather to work more effectively within a chemically centric production model.

University extension services across the nation, quite efficiently, took this information out to real world farmers to help them become better business people and larger consumers of the new chemically driven approach to farming. And over time our country's current industrial agriculture system was born and has flourished.

Now clearly this scenario seems to fly in the face of the Center Garry and Lauren have worked so hard to develop,

so what gives? Garry suggests that the aforementioned 1970s back-to-the-land movement actually had enough of an impact to help organic agriculture get a foothold and allow it to survive. It was significant enough to prompt the publication of extension-type material to help inform people who wanted to move to the country. But it wasn't strong enough to influence most university programs or extension agents themselves.

"The movement to small, organic farms was something a typical extension agent just couldn't support," said Garry. "But the need for information persisted, and a few of us inside the university continually pushed to address it. By the mid-90s we began to develop an organization that focused on small-scale operations, and finally around 2004, what we could actually call the small farms program finally emerged."

Garry's persistence had resulted in the OSU extension service creating a new position description for agents who were specialists in small farms, and even more importantly, were dedicated to organic and biological approaches to agriculture. Ultimately, people were hired to support this new small farms program in five different regions of the state. Plus, additional OSU staff located on the Corvallis campus began to get more involved. Lauren was one of those people.

Lauren was interested in working with the small farms group from the very beginning of her time at OSU. She got involved with the annual Small Farms Conference by helping to organize sessions for it and was asked to come in as a speaker for the small farms' flagship endeavor, the Growing Farms curriculum. Much of her early work was focused on food systems, especially in the area of meat and poultry processing. And as a co-founder of the Niche Meat Processor Assistance Network (NMPAN), she could bring a national perspective to the challenge of recreating a local meat processing infrastructure for small and medium sized producers.

Equally important was Lauren's willingness to take on the sometimes arduous task of working with regulatory issues many small farmers face. So she began working on policy issues and got involved with that at the state level where she worked to help bring about changes in poultry processing regulations that enabled small poultry producers to handle some of their own slaughter operations. She also has worked extensively with the Oregon Department of Agriculture.

Bringing her food systems expertise to bear on small farm issues has helped strengthen the overall small farms program as much as the new extension positions, but both Garry and Lauren point to the extension outreach as the critical element that helped launch the small farms program statewide.

"These weren't five full-time extension positions," explains Lauren. "But it was a big breakthrough nonetheless, and it provided the basis for much of the success we've had so far. It has enabled us to work with people in different ways."

Increased involvement within the state's sustainable farming movement enabled the new extension specialists to begin working with non-profits located throughout the state and assisting with curriculum development for new farmers. It also led to a working relationship with leadership organizations like Oregon Tilth.

"We've developed a strategic alliance with Oregon Tilth," said Lauren. "They handle organic certification, and we provide the research and education component of the organic growing process. That's an example of how we believe our new Center might work with non-profit organizations of all types. Many non-profits working at the local level on public health and food system issues can benefit from the research and education tools we have to offer. And we can benefit from a partnership with them because of the access and relationships they have within their communities."

Both Garry and Lauren believe that partnerships with outside organizations will play a vital role as the Center moves forward because university funding always will be limited. Already as a result of the recent recession they've lost a couple field positions, although they still maintain one field staff member in the central Willamette Valley, another in Southern Oregon who helped write the curriculum for Rogue Farm Corps, and one in the Portland area who focuses on the burgeoning interest in urban farming.

These are significant gains when one considers the fact that much of this momentum is being generated within a land grant university. It's a situation Garry can see the humor in.

"The moment we live in right now is hilarious," said Garry, "because the building we're sitting in – the department of crop and soil science – houses both the organic growers club and the people who do transgenic gene research. All at the same time."

Lauren points out that there are differing opinions in the building, but for the most part they all enjoy a peaceful co-existence. There are some communication challenges, however.

"There's not always an understanding here that small farms are legitimate commercial entities and important players in Oregon's food economy," said Lauren. "But these are not hobby farms. They're Oregon farmers who have different challenges and different markets, and who have to be savvy in all sorts of ways."

"And even though 'small farms' is in our name and it's our focus," adds Garry, "it's not our goal for every farm to be small. We see a lot of overlap with medium sized farms, which most of us would consider to be the heart and soul of American farming. In fact, it's the traditional family farm

– which often is medium sized – that is actually at greatest risk.

"When we look at agriculture like that," Garry continues, "ideally our agricultural system would have a lot of small, highly specialized, profitable farms. These small farms would overlap with a lot of medium sized profitable farms. And the local and regional food system would reflect that, because a medium sized farm is often too big to sell at a farmers market. They need larger markets. Our food system needs to be organized in a way that reflects this and accommodates the different types of producers. A highly developed local food system will provide a lot of important niches and a lot of important products for people. But it also will provide a layer over that which is more regionalized and provides profitable markets for these medium sized family farms as well."

Lauren continues… "And somebody has to grow the raw agricultural commodities that need to be processed into other things. We certainly understand that. So I would say that there are all of those different pieces of it. We think a lot about regional food systems, and as Garry said, profitable small farms are an essential part of local food systems, and they can feed into regional food systems by aggregating or maybe by simply scaling up. We don't have all the answers, but we need to get to the answers. There are a lot of people working on these issues in different parts of the country. We're looking at it from an Oregon point of view."

When looking at the need for a more regional approach to food delivery systems, Garry departs somewhat from the notion that all food must be completely local. He feels that a large segment of the local food movement wants everything to be one hundred percent local. But he sees vulnerability in

The popular BUFA program integrates OSU's established Growing Farms: Successful Whole Farm Management Series into hands-on and season long curriculum. BUFA produces vegetables for CSA and restaurant sales.

that approach, and points to the recent nuclear catastrophe in Japan as an example.

"If everything is local and suddenly it's all poisoned, where are you?" Garry said. "I agree there's a critical need for a local focus, but when we start thinking about the combination of profitability for family farms and an efficient, equitable food system, there's a real need for a regional system and even a national system to back it up. But putting the technical stuff aside, what needs to be more just in this country is the ability for families to farm and still send their kids to college and maintain a good quality of life economically. We're beginning to see more opportunities for very small farms, but we're losing too many medium sized family farms to the corporate operations that just keep getting larger and larger all the time. And that's not the right direction for a sustainable food system to go at all."

It's clear that Garry and Lauren are passionate about creating new opportunities for beginning farmers, rebuilding a local and regional food system, and saving America's traditional, diversified family farms. But one has to wonder how difficult that job might be if they remain inside a land grant university that still puts most of its money and clout behind the chemical/industrial food system that's striving to control the overall food economy. Garry is clear in his response.

"We would like to keep it within the OSU framework if we can. Because by establishing the Center, we already have begun to change the university. If we can create something larger within the university, then we will continue to change the university. And, hopefully, create a model for other universities."

(Clockwise from top left) BUFA program participant Dave Kimball weeds spinach. Stacy Holtmann and Rodrigo Corona harvest Purple Haze carrots for a food donation. Edwin Young readies vegetables for BUFA CSA share packing.

ABOUT

OSU CENTER FOR SMALL FARMS AND COMMUNITY FOOD SYSTEMS
Garry Stephenson & Lauren Gwin
Corvallis, Oregon

Garry Stephenson and Lauren Gwin are the director and associate director, respectively, of Oregon State University's Center for Small Farms and Community Food Systems. The Center is an outgrowth of the OSU extension service's small farms program, which Garry coordinates. It expands the program's work with small farms production and marketing to provide a platform for collaboration across OSU and throughout Oregon, which will help the Center support farmers and help build strong local and regional food systems.

The Center conducts research and provides education on sustainable farming methods, alternative markets, and public policy. It also works to build partnerships with local non-profit organizations focused on issues related to local food and community health.

For more information about the OSU Small Farms Program, please visit the website: smallfarms.oregonstate.edu

Barking Moon Farm

Just west of Applegate, Oregon, Barking Moon Farm is tucked into a clearing on the edge of the Siskiyou Mountains. Here, Josh Cohen is transforming challenging conditions into a highly productive farm.

There have been a lot of changes at Barking Moon Farm over the past seven years, but Josh Cohen is convinced – well, maybe not convinced… let's say hopeful – that everything is finally on track and headed in a very good direction. Although that's not to say the farm hasn't been successful during its first seven years of operation. It's more a matter of reshaping things and staying true to the dreams of everyone involved.

In 2006 Josh and his wife, Melissa Matthewson, bought their property in Oregon's Josephine County. It's a picturesque spot, sitting at 1,800 feet where the Applegate Valley begins to lift into the northern end of the Siskiyou Mountains. Josh admits it may not be the best production land in the area, but it's beautiful, and it's the place where their kids were born. It's home. And it's the place where Josh and Melissa took on the challenge of organic farming.

Josh was coming to farming from ecology and landscaping work. Melissa had just finished graduate work studying sustainable agriculture. And both had completed an internship at another Applegate Valley organic farm. In other words, they had a good idea of what they were getting into, but that didn't make things easy.

"I think just starting with nothing and having nothing to improve was the hardest thing we faced," Josh said. "That nothingness. Trying to go a long ways with some spit and a paper clip, basically. And also just learning this site. Even though we had interned less than fifteen miles away, it's totally different at this elevation. A world different. It's almost like a zone colder here than the fields we lease just six miles down

the road. That's one thing about the Siskiyous… it's really diverse in every way possible."

But start they did. In fact, Josh says they were able to hit the ground running because they were contributing to a cooperative CSA comprised of a group of Siskiyou farms. By only having to contribute rather than manage their own CSA, Josh and Melissa were able to test their property and their infrastructure and figure out what they needed to do moving forward.

Their first year of farming, they planted about an acre of land, which sounds small, but when one considers they both were working full-time off the farm while they tried to remodel a house that had been abandoned for a couple years, an acre was plenty. "We should have bulldozed the house," said Josh, "but we were stubborn, and now it's still a work in progress. We also had our first child that year." That's a lot for one year, but they made it through.

By their second year they were near two acres in production and had begun selling at farmers markets and to a few local restaurants. Josh was farming full-time, while Melissa helped with the farming and worked as an extension agent for Oregon State University's Small Farms Program. Good progress was being made, and when – four or five years into this adventure – Josh and Melissa saw what they thought were some good opportunities to grow and get into wholesale, they made the leap and expanded to a dozen acres of vegetable production. That's the point where things began to unravel.

"The market was there, but we just weren't ready for

These fields, which are leased by Barking Moon, sit beside Highway 238 about six miles east of the home farm. Josh Cohen intensively farms this rich soil as part of an overall effort to keep his acreage small and his productivity high. He's currently experimenting with gradual production increases that include planting one million carrots in preparation for his winter CSA and restaurant markets. That's a 400,000 carrot increase over the previous year. He mentions that Barking Moon is known for it's sweet winter carrots. "It gets much colder here than in California, which sweetens the carrots by turning their starch into sugar."

it," explains Josh. "The interesting thing was that we knew by mid-season that it wasn't working, but we couldn't really figure out why until we looked back on it retrospectively at the end of the season. We needed a better land base. We didn't have the infrastructure. We were selling our best produce too cheaply, which just didn't make sense. All that contributed to make it a very hard year, but it was one of our biggest learning years, so I'm grateful for that experience. I don't think we could have gotten to where we are now without that. But at that point, it was a matter of going back and finding the sweet spot where everything had been working previously."

As it turned out, finding the sweet spot entailed much

more than reducing their acreage and getting out of wholesale. It also meant rethinking priorities and just being honest about their life goals.

After giving farming a real shot for nearly five years, Melissa had come to terms with the fact that she didn't really want to farm. Farming was Josh's dream. Josh says it was always that way, from their first backyard garden. She loved living on the farm, being a part of it and close enough to touch it, but her real passion was writing. A thing she always had done on the side, but never a pursuit she had truly given herself to. It was time to do that. So Melissa enrolled in a master's writing program, and Josh took full control of the farming.

The first thing he did was cut his acreage in half. From twelve acres down to six. Then he winnowed out all of the wholesalers and decided to concentrate on direct to consumer sales through farmers markets and his winter CSA, while continuing to participate in the cooperative CSA and maintaining his restaurant accounts. The turn-around was stunning.

"I thought that scaling back to about half the acreage that our sales might be a little bit less," explained Josh, "but I believed our net would be a little higher, or at least proportionally higher. It turned out that by being really efficient on a small space, our sales were higher than the year before – on half the acreage – and our net was just through the roof. We finally figured out how to make money within our little business. Which is key, because I want to do this every year."

If you're having trouble keeping track of the timeline for this turn-around, the big year was last year. 2013. And now in 2014, Barking Moon is up another thirty percent just by repeating the same processes it instituted last year. With the success he's having, Josh is considering shrinking even more while trying to further refine his efficiencies. But there's got to be a point where that just doesn't work anymore, and he admits that he's still trying to find the balance. The point where size and efficiency combine to yield the maximum return.

In the meantime, he now has more revenue to pay his employees, which means he's able to keep a higher quality year-round staff. And that translates into more time for him to spend with his family and more money to pay for family needs.

"It's hard to do this year after year and to continually do without," said Josh. "Like not be able to buy kids clothes. Take a vacation. Have time to be with my kids. Decide if I'm going to bathe, or eat, or spend time with my family, because I can't do it all. But with the way things are going at the moment, the skill of our employees is alleviating a lot of the pressure that was on me. We even bought a canoe recently, and believe me, that's been a big thing… just going to the lake and cooling down on the weekends."

As Josh and Melissa have struggled and learned and grown through their first seven years of farming, Josh feels like they've learned a great deal – both about what to do and what not to do. So I asked him what advice he would give other folks who are thinking about giving farming a try.

"We've definitely learned that it's important to have a partner who supports you," he said. "Your partner doesn't necessarily have to help you farm, but they have to be supportive of what you're trying to do. And you have to return that support. You both need to be happy and fulfilled or in the long run, things just won't work because there's too much pressure.

"We've also learned that getting bigger is not necessarily the path to more money or a better return or a better quality of life. You don't have to stay small to succeed, but you have to make the choices that work best for your operation. Like we discovered that wholesale doesn't work for us. It actually puts a lot of small farmers out of business. That's probably what would have happened to us if we tried to stay with it, but it was important to try it. You have to try everything to find out what works. Different markets, different crops, different varieties. That's how you find out what works best for you."

Another point that Josh makes echoes the comments I've heard from so many farmers… first and foremost, if you're going to choose farming, make sure you love it. To help you figure that out, Josh recommends doing an internship. You could just get a job as a farm laborer, and you no doubt would learn a great deal over time. But internships provide more of a big picture educational component. You get a glimpse of the many different aspects of farming that just working in the fields doesn't provide.

Barking Moon Farm typically accepts one intern each year from Rogue Farm Corps. According to its website, RFC's programs combine hands-on training, classroom learning, and farm-based experience on family farms in Oregon's Rogue Valley and the southern part of the Willamette Valley. ATTRA, the National Sustainable Agriculture Information Service is another source for both people wishing to be interns and farmers seeking help.

All of the tasks involved in farming, the growing, the marketing, the bookkeeping, managing employees, acquiring and maintaining equipment, it's a lot to wrap your head around when you're just getting started. That's why Josh feels that any new farmer will have best chance for success if he or she starts small.

"It's so important to start really small," he said. "Partly because there's so much to learn, but also because that makes it easier to diversify your income by having some sort of off-farm job, at least in the beginning. Farming isn't going to provide a return very quickly, and diversity helps manage risk. Income diversity, as well as diversity in your markets and your products. What we've found really helps us is spreading things out. Unfortunately, this approach means there are no major wins, but there also are no major losses."

In other words, stability translates into sustainability. And sustainability means making enough money to pay the bills.

"You know, a lot of people talk about being sustainable in terms of farming practices. And we do, too, but for us, first and foremost, we base our decisions on a model of financial sustainability. Because we want this to last. We love being here, and we love doing it. Money's a real part of that, and we quickly found that out. So for us what feels most comfortable is being able to pay the bills. I don't have to be a rich guy, but we still have debt from our start-up, and I don't like owing people money. It makes me uneasy if we can't come through and pay our bills. So having more coming in than we have going out is what feels sustainable to us."

A *Barking Moon Farm intern does some online work during a hot summer afternoon. Barking Moon usually hires one intern each year, selecting the candidate from the Rogue Farm Corps program. Rogue Farm Corps offers internship participants a well-rounded program that includes both classroom training and actual experience on farms in the Rogue Valley, Applegate Valley, and southern Willamette Valley.*

Listening to Josh makes it clear that farming's not for everyone. He believes that what gets a lot of people into it is all the romantic notions, and he admits that he never thought his foray into farming would lead him to this farm. "Initially, I kind of thought maybe we'd be weekend warriors. We'd do a farmers market with a couple heads of lettuce. My notions were sitting on the deck with a cup of coffee in the morning looking down and saying that everything looks good."

He didn't realize he would be an employer or doing payroll or negotiating leases. In fact, Josh says that much of the time he doesn't even think of himself as a farmer. He sees himself as a small businessman who happens to be doing farming. But regardless, farming is what he definitely wants to continue to do for as long as he can.

Josh grows thoughtful as he projects his thoughts into the future… "I think I want to live here forever… on this particular property. I just love it here. Both our kids were born here, and I feel like they want to live forever, too, or so they say right now. But they're three and seven. But I am a little frightened to lock myself in and say that I'm going to be a farmer forever. Our business may change later in life. Just as I get older. I don't really know what will happen down the road, but for now I'm trying to take care of myself so that I can do this for a long time."

(Clockwise from top left) Applegate's general store and cafe. Barking Moon farmhouse sits at the west end of the home property. Fields along the Applegate River.

ABOUT

BARKING MOON FARM
Josh Cohen
Applegate, Oregon

Barking Moon Farm is located in southern Oregon's Applegate Valley about halfway between Grants Pass and Medford. The farm got its start in 2006 when the property was purchased by Josh Cohen and Melissa Matthewson, and its first crops were produced in 2007.

Barking Moon produces a diverse mix of organically certified vegetables, legumes, and grain crops. It primarily sells direct to consumers at farmers markets, a December to February winter CSA program, and as a contributing farm to the Siskiyou Sustainable Cooperative CSA that runs from June to November. It also provides organic produce to area restaurants.

For more information about Barking Moon Farm, please visit the website at barkingmoonfarm.com.

Minto Island Growers

Elizabeth Miller and Chris Jenkins have turned Minto Island Growers into a multifaceted farm that features a tea plantation, vegetable CSA, u-pick berries, food cart, farm stand, and more.

The scene when we pulled up to the Minto Island Growers' farm stand was one of happy activity. A food cart was prepping meals, several picnic tables were filled with diners, a couple of farm hands were planting and watering flower starts, and a cashier was adding up the total for a woman buying vegetables. I was convinced we had arrived at a successful farm, because there was simply too much going on for it not to be so.

Chris Jenkins and Elizabeth Miller are the farmers making all this activity possible. They've been farming for seven years now, and they've jumped wholeheartedly into virtually every opportunity that's come their way. In fact, there's so much going on that I wonder how they manage to get it all done… and whether the workload has any impact on their outlook.

When asked, both express confidence they'll be doing this work for the rest of their lives, and when they take the time to think about it, they're pretty sure they still love it. But they are tired. Mentally, physically, emotionally. Seven years of running as fast as they can to take advantage of every door that opened has taken a toll. And now, as we sit at a small wooden table in the corner of a metal warehouse large enough to park a few trucks in, they almost seem relieved to have a reason to sit down and reflect for awhile.

When they got started as Minto Island Growers back in 2008, they were carrying the ideas and ideals they had developed during some fairly radical ecology and economic

studies abroad coupled with a one-season internship at a free-thinking California farm. But it's not like they were coming into this adventure blind. Elizabeth grew up on this farm, and in her mind, she had never really left. She came back most summers during college to help out her dad, who specialized in native plants, traditional poplar breeding, and mint leaf production. In a sense, Elizabeth's path to having her own farm was more like taking a sabbatical for a couple of years, during which she found her own style and developed a context for agriculture that came from the world beyond her father's farm.

Though Chris didn't have any direct ag experience growing up, he spent as much time as possible exploring the Ohio countryside of his family's 150-acre country home, where his physician father escaped the pressures of his profession. Chris became enamored with nature to the point of eventually studying biogeochemistry in college, doing urban ecology work in New York City, and working for a California-based restoration and living architecture firm that used plants functionally within the built environment. But as he found himself spending more time in front of a computer than outdoors, he knew he needed to come at his love of plants from a different direction.

It was in this context that Elizabeth and Chris came together to spend that single season at the California farm doing something completely new to both of them. Growing

Growing vegetables for the Salem, Oregon market is one of Chris and Elizabeth's favorite activities.

vegetables. They fell in love with it. And vegetables, it turns out, more or less defined the launch of their own farming operation.

"When we got started here, there were opportunities for us to carry on some of the work my dad was moving away from, like the native plant nursery," said Elizabeth. "But we were more passionate about vegetable production. We both just love growing vegetables. And especially for me, it was also about bringing to Salem, which is my hometown, the community value that fresh organic vegetables represent. Organics are thriving in Portland and Corvallis and Eugene, but there aren't a lot of organic farms in this area. And here we are with this large land mass ten minutes from downtown. We saw this farm as a huge community asset just waiting to be developed."

It didn't take long for Chris and Elizabeth to get a vegetable-based CSA up and running. As they were kicking off their own CSA, Winter Green Farm, an established

biodynamic farm located in Noti, was wanting to eliminate it's Salem drop. Picking up that drop enabled Minto Island to begin with a substantial foundation they've just continued to build on.

Part of what they've been building is the farm's relationship with the community of Salem. To help with that, they added a farm stand where locals could pick up a CSA share or simply buy vegetables and fruit. The stand also provided a base for the farm's u-pick blueberry operation. And though the farm stand was performing okay on its own, Chris and Elizabeth felt it needed a bigger drawing factor, so they decided to add a food cart that could provide meals for anyone wishing to come out to the farm.

From CSA to u-pick to farm stand to food cart. It seems like these young farmers haven't been able to stop themselves from continually looking for ways to make each aspect of their operation more successful. And this is all on top of the "other opportunities" they took over from Elizabeth's father, which included a native plant nursery that supplies various environment restoration projects, a poplar breeding operation that provides plant stock to an eastern Oregon timber operation, and a business growing mint plugs for Oregon mint leaf producers.

"To be honest, a sense of responsibility has driven a lot of what I've done on the farm," said Elizabeth, "but I'm starting to understand that no matter how much of a strong ethic and value system there is behind a dream, you have to be able to live a balanced life in order to fulfill that dream. And that's very challenging in agriculture."

"And there's also the tea," injected Chris. At which point Elizabeth visibly drooped. In recounting for me the multitude of businesses they're trying to juggle, she had forgotten the one that is currently generating the most buzz and about which they both are most excited. Half an acre of 24-year-old tea plants.

"Elizabeth's father, Rob, planted the tea in the late 80s with a partner," explained Chris. "Not much happened with it initially, but we're now working with a processor from Chehalis, Washington. He went to China to learn a very specific style of oolong processing at a village there and is really passionate about it. We've been through a number of trials and now we've got a product we feel like we can really stand behind. But, of course, developing this marketing opportunity doesn't do much to solve the problem of doing too much."

Elizabeth adds, "The economics of the tea project are really challenging because all the tea leaves have to be picked by hand, and there's no cultural expertise here for that. On the other hand, this opportunity is simply too unique and too promising to walk away from. So the logistics and the work load is something we'll have to continue to work on."

Our Minto Island duo admit they already have cut back on some of the native plant and timber-oriented operations they originally inherited. But because those endeavors always were the least labor intensive, their load hasn't lightened all that much. And they are realizing that labor – the cost of it – will be one of the primary factors that shapes the way their farm evolves.

What many people outside the farming industry fail to understand is how much CSAs depend on low-cost labor. For farmers just starting out with a few acres, a little training, and a dream, labor cost is not a priority because they're happy to simply recoup the cost of their seeds and inputs, put food on the table, and make their lease payment. They're doing the work themselves and most realize going in that they're not going to be able to pay themselves a salary during their early years. Growth changes the equation. Labor and the true cost of doing business have an ever increasing influence on how a farm operation evolves.

"In order to keep all these things afloat and have it make sense – and I do think it can make sense – we need to find the right combination of factors that will keep key employees committed to the farm long-term… people who can help manage each component of our operation," said Chris. "We've had great help from family, certainly. And several key employees have been here awhile now and helped manage things, but they're moving on, and when they do, it will be difficult to start that cycle over again. We just don't have enough time to do everything."

"We can't do it all and do it well," added Elizabeth. "We're not managing things as well as we could, so we're losing money and losing time. We make it work. Honestly, I think we do an amazing job for how much we have going on. But we're feeling a tinge of burnout, I think."

Chris added, "Sometimes we're reluctant to say that to people because there's a tendency still to romanticize farming a bit and believe that we're living the dream. And a lot of the time we are living that dream, but it's really friggin' hard."

"We do feel blessed," continues Elizabeth. "We're still really stimulated by what we're doing. We're both fascinated by the art of growing a diversified farm. But I think no matter what, even if you have that, if you're doing too much, it just doesn't work at the end of the day."

"I think part of it is that in the beginning it was easier to juggle everything because the volume in all of the sectors was lower," Chris commented. "And each one now – farm stand, food cart, farmers market, CSA – all of them have grown to the point where it's getting harder to manage."

Elizabeth and Chris essentially have been overwhelmed by their own success and need to come to terms with that. So now they're trying to discipline themselves and really keep track of costs. Then attribute those costs to the appropriate sector, because that information is needed to make smart business decisions. Figure out the production costs of mint and just mint. Parse the cost of vegetables, but not just vegetables… vegetables as they relate to CSA costs, farmers market costs, farm stand costs. Plus, they're trying to be sensitive to the human costs of too much pressure.

"When I think about what I love most, it's still growing vegetables," said Elizabeth. "Whenever I'm out on the tractor or looking at the crops, that's when I'm happiest. With the pure art of growing food. That process is incredibly satisfying."

Minto Island Growers is located so close to downtown Salem, the addition of a food cart to its farm stand operation makes sense and increases traffic.

It's that irreplaceable sense of satisfaction that keeps these two earnest and intelligent young people moving forward and focused on their organic farming future.

"We understand that we need to take a hard look at what all this takes," said Chris, "whether it's the marketing or what you're growing or adapting our farming techniques to outside forces. But I'm pretty confident that we have the resilience to adapt. And I guess I don't question whether or not we'll be farming here for the rest of our lives because I've internalized the fact that it's assured."

"I don't think any other work could be as satisfying or as

enriching," added Elizabeth. "It's just a matter of finding a little more balance and making decisions that work for us rather than just doing things because we feel that we have to do them. Which is where we started out. I mean, we're thankful for all the opportunities that have come out way. But I think it's really easy in agriculture to believe that this work is so important that you'll sacrifice anything to make it work. I now realize that is not the truth for me personally. I need personal balance. Because the joy you feel doing the work is reflected in what you produce. I want to bring more of the joy back into it. I think we'll find that."

And with all of the growing pains and challenges they face, Elizabeth and Chris still would encourage other young people to give farming a try.

"I would say don't romanticize it, but don't be afraid to try it," said Chris. "A lot of people are farming organically and sustainably and they're making it work, so it is possible. It's important to recognize whether you're feeling fulfilled, because if you're not, then you're going to run out of energy. But at the same time, it's good to know that there's always next year. And it's amazing and humbling to know that our business is completely at the whim of natural forces. It feels true to the reality of things, of the world. It just feels real." ✕

(Clockwise from top left) Vegetables on display at the farm stand. Part of the extensive infrastructure on the farm. U-pick blueberries are a Northwest favorite.

MINTO ISLAND GROWERS
Elizabeth Miller & Chris Jenkins
Salem, Oregon

Minto Island Growers is a diversified farm and nursery located in South Salem, Oregon. It is co-owned and operated by Chris Jenkins and Elizabeth Miller, who lease land owned by the Miller family. They began farming at this location in 2008 and began the transition to organic fruit and vegetable production that same year. Portions of the farm are certified organic by Oregon Tilth, and the entire property is farmed under the USDA's National Organic Program standards, but Elizabeth and Chris view those standards as just a starting point for their management practices.

Minto Island provides its customers with organic produce through a CSA, farmers markets, and a u-pick berry operation, as well as a farm stand and an on-farm food cart. In addition to their work growing fruit and vegetables, they carry on the work of Elizabeth's father with native plants, poplars, and mint propagation for leaf production.

Elizabeth and Chris are dedicated to making connections with their local community, supporting the local economy, and nurturing the farm ecosystem with ecological growing practices and stewardship.

For more information about Minto Island Growers, please visit their website at mintogrowers.com.

Sweet Home Farms Meats

Sweet Home Farms Meats is located on acreage in the central Willamette Valley that includes a picturesque stream that offers both water for the farm and a great place to cool off on hot summer afternoons.

When I sent an email to Sweet Home Farms owners Mike Polen and Carla Green, I was anticipating talking with a couple of sixty- and fifty-something, doctoral-level healthcare researchers who made a stunning career shift about seven years ago when they left Kaiser Permanente in Portland to buy a ranch near Sweet Home and begin a sustainable livestock operation. Instead, I heard from Mike's son, Daniel O'Malley, who in 2013 took over Mike and Carla's meat production business and was anxious to promote it.

Mike and Carla still run Sweet Home Farms, but they concentrate on breeding and raising belted galloway cattle and katahdin sheep, as well as great pyrenees and English shepherd dogs. The labor-intensive direct-to-consumer meat production business — Sweet Home Farms Meats, LLC — moved over to Daniel. And Daniel clearly loves it.

To say Daniel has come to farming via a different route than his dad would be too much of an understatement. In fact, the route is about as far from Mike and Carla's journey as one could get. Rather than thoroughly researching farming and ranching as an alternative career, Daniel simply needed a job. He was completing an undergraduate degree in political science at the University of Oregon and wasn't sure what to do next. He wasn't worried about it or anything like that… he's pretty much a free spirited guy and doesn't seem to be overly worried about anything really. Which I suppose can come in handy when you're trying to get a farm started. So when

he learned that Mike and Carla needed help, he saw it as an opportunity and jumped right in and spent his last summer before graduation learning to farm and ranch.

"My Dad and his wife, Carla, bought this farm in 2006," Daniel explained. "As I was finishing school, I started coming up here and helping out, and I just really started to like it. So I finished my last term, and then we all talked about turning this into a real business. They had been doing it, but at a pretty low-volume level… it was still similar to a hobby. When I joined, we decided to really push this and went full steam ahead."

Working on the farm was the first job Daniel had out of college, so from a professional standpoint, he's never done anything else, which suits him fine.

"I never knew what I wanted to do growing up, but I knew what I didn't want to do, and farming was not one of the things I didn't want to do. But I definitely did know that I didn't want to work inside five days a week. I just thought it would be numbing. So when this came along, I thought, wow… there's something new everyday. It's interesting. You have to problem solve. Sometimes it's long term problem solving, but other times it's like the cows get out, and if you leave what you're doing right then, the pigs will get out, so what do you do? You've just got to make a decision and go with it, and that's fun."

Daniel is very surprised by how quickly he reached the

The cattle shown out in this pasture are raised using a process that Daniel O'Malley terms Management Intensive Grazing. This is a grazing technique that is growing in popularity because it not only provides a good quality food resource to the cattle, it improves the soil and plant diversity in the pasture. Daniel also rotates different types of animals through his fields, such as cattle followed by chickens, to help prevent parasite build-up in the soil.

point of owning his own farming business. When he began this undertaking, he told his partner, Bryn, who lives with him on the farm and works in Eugene, to give him five years and see what happens. Two years later, he had his own farm. Although he did comment that "it felt like five years, actually. I worked enough hours." But the hours were worth it because he loved working with the animals and loved living on the farm.

Living the life of a farmer has brought other changes to Daniel and Bryn, as well. "We're learning so much more about food. Our diet has changed a lot the past couple of years. Not because that was some kind of goal, but because of being out here, producing the meat, being around all of the other farmers at markets and seeing what they're growing. As you get closer to the place where the food is being produced, you just start eating healthier." But the biggest change for Daniel is simply being in charge of running a business.

The business, itself, is actually a livestock feeding and retail meat operation. He does no breeding. He purchases cattle, pigs, and chickens at a young age, then raises them until they're ready for market and hauls them to Mohawk Valley Meats, a USDA facility located about thirty minutes away in Springfield, Oregon. The lamb he sells he purchases directly from Mike and Carla, and his goat meat is raised by another farmer located down the road a ways.

He is proud of the fact that almost all of the meat he produces originates from within fifty miles of his farm. The one exception to date is that some of his cattle come from central Oregon a couple hundred miles away. He understands that he is somewhat at the mercy of the marketplace. The goal, however, is to keep everything as local as possible, and as his business grows, to partner with other farmers to create a cooperative approach that spreads both the risk and the opportunity throughout the local agricultural community.

These are lofty ambitions for someone who is just getting started, but Daniel believes his motivation, youth, and energy can carry him a long ways. On the other hand, he's realistic about the challenges and risks, and knows everything could change very quickly.

"I lease this nineteen acres from my dad and Carla, so if anything happens to their farm, this place would be in jeopardy," Daniel said. "Also, having access to equipment like my dad's truck and trailer is critical. If I had to buy my own truck and trailer, it would put me out of business. And at this point in my business, if something happened to the animals… if they got sick and I lost them, that would be devastating. But I don't see that being a whole lot different from starting any kind of business. A lot of things can make it not work, but that doesn't mean you shouldn't give it a try."

The isolation of being so far out into the country also has required both Daniel and Bryn to make a major adjustment.

"I grew up in Portland, and Bryn grew up in Seattle," explained Daniel. "And as the months go by, we're realizing more and more that there really just aren't many people out here, and that's kind of hard to deal with sometimes. We occasionally talk about it, but we always end up asking ourselves where else are we going to go that we like better. We do like this place. We love coming down here by the creek. It's beautiful. So as long as we can keep things going, and ultimately growing, we see ourselves being here."

Daniel feels like he always has had an affinity for nature, and he learned holistic management techniques from his dad that should enable him to continually improve the quality of his farm's natural resources. When working for Sweet Home Farms, Daniel learned the practice of Management Intensive Grazing (MIG), which requires the regular movement of livestock from one area of grass to the next. He also learned to rotate different types of animals through the various fields to create positive symbiotic relationships that result in organic soil improvements.

"The land is always doing its thing, but we're trying to get higher levels of production by keeping things holistic and organic," Daniel explained. "When you use synthetics, you actually deplete soil microorganisms and other benefits. So you're actually losing something… it's a false gain. Whereas what I'm trying to do is make it a real gain. It's still a little early to tell on this farm, but over at my dad's property you can easily see how much more grass there is… how much more variety of plant types there are. Their dry matter per acre has to be five or six times what it is on neighboring farms that use synthetics. A cow can be on that acre five times longer, which means we can put extra weight on the animal. So we just made money by improving the land naturally."

Daniel went on to explain that natural soil improvements begins with grass, which grows in an "s" curve – a slow-growing first stage, a fast-growing second stage, and a slow-growing third stage, which includes the seed production that leads to dormancy. If a grazing animal eats the grass during its fast-growing stage, the grass goes back to its first stage because it can't photosynthesize as well and the roots take a little shock. Some roots are actually killed. When those roots die, small air pockets created. Then, if the grass is allowed to return to stage two growing, new roots will retake those air pockets and also make new ones. Repeating this process over and over results in aerated soil. In turn, aeration results in better water retention and allows both more grass and other plant types to grow.

The key is to avoid allowing the grass to be bitten into a second time before it grows back to stage two, which is accomplished by moving the cattle from one paddock to another every two to three days. And following the cattle with chickens also helps by providing a rich natural fertilizer. But even fertilizer additions need to be halted until the manure can adequately decompose before more manure is added to the grass.

Daniel's explanation makes it easy to understand why the technique is called "management intensive." And the quantity of information and level of detail that Daniel readily and easily imparts is impressive knowing that he's only farmed for a couple of years.

"I think you should do what works for the long term,"

said Daniel. "And all that begins by improving the soil, increasing plant diversity, and raising better grass. I'm basically a grass farmer. But I know I've got a lot to learn. Like I don't really know all the different types of grass, and I don't feel a need to run out there and test the protein levels of all the different grasses. I figure if my grass starts growing earlier this year than it did last year, then that's good enough for me. You don't need any science to see that grass gets thicker and darker green after you run chickens through it."

Though Daniel's partner, Bryn, has her own business which requires a full-time commitment in nearby Corvallis, having both her and Daniel's mother available to assist with farmers markets and CSA drops from time to time provides a significant benefit. As any beginning farmer knows, the need to hire help can create a major obstacle to having a profitable farm.

Everything Daniel conveys makes sense from a long-term business standpoint, but it's clearly a risk when one considers that his land is currently being leased rather than purchased. When I point that out, his tone becomes more thoughtful, but he remains steadfast in his opinions.

"Leases come and go," he said. "I feel pretty secure here right now, but I am setting up a business while remaining fully aware of these variables. That's partly why I'm only feeding animals right now rather than breeding them, because I can adjust what I have at any time. I can either buy more or sell what I have. If I had a cow-calf operation and I lost a lease,

I'd have a really hard decision to make. So I'm not going to get into that. Or at least I won't do that until I have enough reserves to be able to afford it. Until then, I think it's important to stay very flexible."

Daniel believes that flexibility and creativity are the keys to successful farming. And that planning is good, but those plans need to stay loose.

"I think farming is as much or more an art as it is a science. You've got to just go with it. And hopefully at the end you've got something really good. But as you're going, you don't know what's going to happen. I'm trying to let what's around me, everything that happens, guide me, and help me understand what's possible.

"I believe life's a river, and I should be flowing in that river. That doesn't mean I just close my eyes and hop on a tube and go wherever I want, but I should be flowing in the right direction. There is energy moving all around me, and I should be moving with that energy. If you try to fight it, you'll run into all types of problems. So I try to approach farming that way. Just go with it, and observe, and see and feel where things need to be moving."

(Clockwise from top left) A few chickens gather to check things out. Berkshire hogs are a heritage breed. Sheep are also raised on the farm.

ABOUT

SWEET HOME FARMS MEATS
Daniel O'Malley
Sweet Home, Oregon

Sweet Home Farms Meats was established in 2013 when owner Daniel O'Malley took over the meat production and sales business from Sweet Home Farms, which is still owned and operated by Daniel's father and step-mother, Mike Polen and Carla Green.

Sweet Home Farms Meats raises and sources animals that have been raised in ways that mimic their natural habitat and encourage natural behaviors. The farm practices low-stress, management intensive and multi-species rotational grazing.

Grass fed and finished beef and katahdin lamb, pasture raised poultry, and humanely raised pork are offered in CSA shares, bulk purchases, and individual cuts through the farm's website and at farmers markets in Corvallis and Portland.

For more information about Sweet Home Farms Meats, please visit the website at swmeats.com.

Greenwillow Grains

In the heart of Oregon's conventional grass seed production region, Willow Coberly has successfully worked, campaigned, and collaborated her way to a new local grain milling operation.

Anyone who pays attention eventually learns that things aren't always what they initially appear to be. A case in point is Stalford Seed Farms in Tangent, Oregon. At first glance, this farm appears to simply be one more large-scale conventional grass seed farm, for which Oregon's Willamette Valley has become well known. But a closer look shows us it's so much more than that.

Owners Willow Coberly and Harry Stalford do raise more grass seed than anything else on the six thousand acres they farm, but they also farm more organic acreage than most other farms in the valley as well. But they didn't always.

Willow had long wanted her husband and business partner, Harry Stalford, to grow food rather than just grass seed. Plus, Willow wanted the food being grown to be organic. She was a cancer survivor herself, and she was seeing cancer show up in so many people around her. Many of these people were systematically exposed to the heavy dozes of toxic chemicals used in conventional farming, so she wanted the farm ground around their house to be free of chemical pesticides and herbicides.

"Basically, I wanted to keep my kids safe," said Willow. "I looked around, and it seemed like all of our neighbors were affected by cancer in some way. I believe all these chemicals contribute to that, and I didn't want it near my house. When I first started talking about this, people didn't take me seriously. People like Harry's mentor, who was an old guy, seventy-something years old, just laughed about it. He was the kind

of guy who mixed chemicals by hand while he smoked a cigarette. He never got cancer, so it was hard to argue with that anecdotal evidence. But then the guy just died suddenly, and he had cancer all through his body.

"Even after that happened, I still had a hard time convincing Harry to stop spraying until I asked him about his own son… that if his son was laying in bed dying of cancer, would he be able to look him in the eye and honestly say he had done everything possible to prevent it. That's what finally got his attention and made him agree to stop spraying the chemicals. Getting those acres certified organic was a different challenge altogether, but it was a start."

The road from those first poison-free acres of farmland to Greenwillow Grains was a challenging one which, Willow explains, relied on the concerted efforts of a number of different people. But Willow was confident from the very beginning that the successful conversion of conventional land to organic was possible, especially after she and her mother, Gian Mercurio, attended the 2004 International Federation of Organic Agriculture Movements (IFOAM) meetings in Rome.

That time frame was important to Oregon's burgeoning new farm movement because the level of activity began growing rapidly, with new projects taking shape throughout the Willamette Valley. One of those projects, which was aimed at increasing food security, resulted in the formation of the Ten Rivers Food Web. From a farming standpoint, one of the key developments this new organization achieved was bringing

Following harvest, this field of organically grown oats will be milled and packaged at Greenwillow Grains. Owner Willow Coberly's goal is to grow the local market for her mill's rolled oats and wheat flours to the point where all of her products are consumed within a fifty mile radius. Until then, she will continue to ship to customers around the country.

both conventional and organic farmers together to talk about food security and the possibility of converting some acreage from grass seed production back to food production.

Willow and her husband Harry were part of those conversations, and before long Willow ended up joining the Ten Rivers board of directors. Willow and Harry already had converted about 135 acres to organic food production and were successfully growing five varieties of beans on a commercial scale. While the beans were growing, Ten Rivers applied for and received a grant to see if, after harvest, Willow and Harry's commercial seed warehouse could successfully clean and process the dry beans.

According to Willow, Dan Sundseth and Harry MacCormack, both founding members of Ten Rivers, were integral in the early formation of both the Ten Rivers organization and this bean production project, which came to be known as the Southern Willamette Valley Bean and Grain Project.

The Bean and Grain Project aimed to increase local food security by trying to close the loop in the process of growing staple foods in the Willamette Valley, and keeping those foods in the Valley rather than shipping them out. Naturally, since the goal was healthy food, the bean and grain products would need to be grown organically, which also served to increase organic production acreage.

The early bean experiments were an immediate success, and now other growers have begun participating in bean production. But most of the local farming community remained skeptical about the 'grain' part of the bean and grain project.

Most wheat grown in the Willamette Valley has always been white wheat, which is used in pastry or pasta flours. So there was some consensus regarding white wheat. But bread making requires the higher levels of protein found in hard red wheat, which typically doesn't grow well in cool, wet conditions.

True food security, however, requires bread, so Willow rose to meet the challenge head on. She does admit that in the beginning, she was having a hard time finding any red wheat seed to try. But eventually she was directed to a researcher at the University of California-Davis who had been working on a type of red wheat, which she was breeding for cool, wet conditions… just what Willow was looking for.

"I called her up and told her I had heard she had a red wheat that would grow in a cold, wet climate," recounted Willow. "And she said it was interesting that I called on that particular day because she was literally packing up her office because her program was being shut down. But she had 25 pounds of her wheat sitting there and was wondering what to do with it. So she agreed to send it to me as long as I promised to set aside a small amount in its pure form. So she mailed me that seed stock, and we kept five pounds back."

The remaining twenty pounds were mixed with some other seed stocks Willow had acquired from Montana and North Dakota, giving Harry and Willow about 100 pounds of seed wheat to plant that first year. The total harvest from their initial planting was only 75 pounds of grain.

"You can imagine what kind of reaction that got," said Willow. "But I said, let's plant this 75 pounds and see what happens. Harry just laughed, but we did it, and the harvest from that 75 pounds was 3,000 pounds of grain. Now, most of the red wheat growing in the valley comes from that 3,000 pounds. We've never changed it. We just save our own seed, which is another thing we were told by the big seed companies would never work. They said we needed to buy new seed every year, but where is that new seed supposed to come from?"

The next lesson Willow learned is that growing the grain is only part of the process. Once grown, the next step is selling it. Unfortunately, the processing infrastructure for milling the grain into flour also had disappeared from the Willamette Valley. So initially Greenwillow Grains tried selling whole wheat berries, encouraging customers to mill it themselves, which really didn't work. So they decided to get a small mill and grind some flour so people could taste it. Then a bakery in Cottage Grove, Oregon became their first customer, which meant that the business of milling flour from locally grown grains had finally returned to the Willamette Valley. Thanks largely to the persistence of Willow Coberly.

Greenwillow Grains has grown substantially from this meager beginning. It now has five different mills of varying sizes used for milling different products. It also has a capable manager – a neighboring farmer named Mike Robinson, who lost his farm while attempting to shift it from conventional to organic growing and was looking for a way to stay involved with sustainable agriculture.

"Mike losing his farm was a big part of this," admits Willow. "He transitioned into running the mill, and he's been the best thing that ever happened to Greenwillow Grains."

Mike mills Greenwillow's bread and pastry flours, and Willow's son, Christopher Richter, handles the oat rolling. It's the oats which are proving to be Greenwillow's specialty.

"In much the same way that Willow came up with her red wheat seed stock, I think we kind of struck gold with the original oat seed we got," said Mike. "It tastes great, and it grows well here. It doesn't seem like there are many similar varieties around here. Seems kind of rare. And from our standpoint, I hope it stays that way, because it produces a wonderful product."

"Oats are really our niche," continued Willow, "but I'd put our pastry and cake flour up against anyone's. I do a lot of baking, and they're really good. But the oats are special."

Greenwillow's founding and expansion has been slow and steady, coming over a number of years of trial and error. But what has expanded even more slowly is the local market, somewhat to Willow's consternation.

"To be honest, we thought there would be a bigger local demand," explained Willow. "I wanted a customer base that lived within a hundred miles of us. Instead, we have customers on a Pacific island, a few on the East Coast, some in the Southwest, and even Wisconsin. I don't know these people found us. I mean we have a website now, but we didn't early on. They just got our number somehow and called us and started ordering. It's great having them as customers, but what I'd like even more is for them to get their own mill and community together and produce their food locally."

Willow plans to continue shipping products to her distant customers for now because that revenue is helping Greenwillow Grains pay its own way. But she's hopeful the Willamette Valley market will eventually expand to the point where all of her customers are located nearby. That was her original vision, and it's the one she continues to hang on to.

She explains that for her, a business should strengthen itself from the inside out. First take care of its employees, which Greenwillow does in a variety of ways, including giving them whatever they need from the company's production and by providing them with an organic community garden. Rather than saying to employees, this is what the market will bear so I'm only going to pay you three dollars an hour, everything starts with paying a living wage and providing healthcare to your employees, and you work out from there. The next step for a company is to provide for the needs of its community, and finally it should market to the wider region.

"I always will take care of the Corvallis co-op first," she said. "Then we'll take the next step out. But it starts with employees and your neighbors. We're doing this to provide safe, quality food to our local community… to provide more

Willow and mill manager Mike Robinson, stand outside the original Greenwillow Grains mill. An expanded milling space with larger equipment is currently under construction in another building adjacent to this one.

food security and to create good jobs right here. The people who do the work need to be paid fairly. I just don't believe exploiting farm workers, either here or in third world countries, is the right thing to do."

Willow's husband Harry has also learned a lot as they've gone through this process. Especially about the benefits of taking better care of the soil by using more organic methods. Though he struggled against it in the beginning, he now acknowledges that an organic approach improves the soil and leads to better crops. And when we spoke briefly about farming methods, he even suggested that his soils would benefit from better rotation and rest, but admitted that actually doing that isn't as easy as it first appears.

"A lot of fields… soils… would improve if they were allowed to lie fallow on a regular basis," he explained, "but it's not always up to us whether or not we can do that. I might want to do it to improve the soil, but if the bank owns the loan and the bank says to grow a crop, I have to grow a crop."

What Harry does have control over is how that crop gets grown. Today, his methods, even on his conventional fields, are far more organically informed that ever before. For example, he has learned that he can reduce chemical dependence (and costs) by spraying the soil with compost tea, an organic soil improvement process that spreads the living organisms in compost throughout the soil.

Clearly, the process that led to Greenwillow Grains continues to unfold, rippling through the lives of everyone involved like a Willamette Valley breeze blowing through a field of hard red wheat. ✕

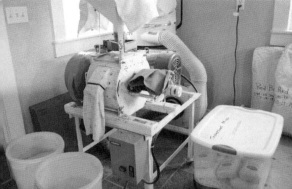
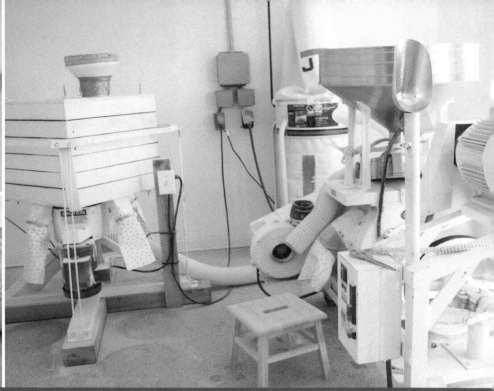

(Clockwise from top left) Greenwillow Grains products on display in a Brownsville, Oregon coffee shop. Milling equipment in the expansion facility. One of the original mills.

ABOUT

GREENWILLOW GRAINS
Willow Coberly
Brownsville, Oregon

Greenwillow Grains is an organic grain mill located in Brownsville, Oregon, that specializes in stone-ground flours and rolled oats. The company is owned and operated by Willow Coberly.

All the grains used to produce Greenwillow's products are grown organically in Oregon's Willamette Valley. In addition to the goal of local production of all the grains the company mills, it focuses on expanding the number of acres in the valley dedicated to organic growing methods and strengthening the local economy by creating local jobs and supporting local markets.

In addition to producing its grains locally, Greenwillow uses local resources for all of its needs whenever possible, and sources milling equipment, bags, and cleaning and packaging equipment from U.S. manufacturers.

Greenwillow's product line includes the flours and berries of hard red wheat, soft white wheat, and triticale, as well as their extremely popular rolled oats.

For more information about Greenwillow Grains, visit their website at: greenwillowgrains.com.

Friends of Family Farmers

As a high profile advocate for and supporter of organic and sustainable farming in Oregon, Friends of Family Farmers is involved in everything from education to legislation that supports small farmers.

Just to be clear, Nellie McAdams is a farm kid, even though she spent the majority of her growing up time in Portland. Every weekend found her on the family's hazelnut farm near Gaston, Oregon. Because that's the way her parents farmed. They worked office jobs in the city during the week and farmed on weekends, amounting to two full-time jobs each. Years later, that's still how Nellie farms today.

Each weekend she drives from her home in Portland to Gaston to continue her agricultural apprenticeship with her father, clearly with the hope of one day operating the family hazelnut farm herself. And now that her dad has turned 69, that future is fast approaching.

But being a weekend only farmer doesn't bother Nellie a bit… quite the contrary, in fact. She prefers it that way. "My parents never really made that one or the other kind of a choice," she said. "I love farm work, but I also crave the social advocacy piece. So ideally, I'll continue to do both."

Which brings us to the social advocacy piece. When Nellie and her dedicated co-workers aren't farming, they help others farm.

Developing and implementing programs for Friends of Family Farmers, or FoFF, is how Nellie spends the majority of her time. Her primary tasks are overseeing the organization's "next generation" efforts, but as with most small non-profits, the hat she wears depends upon the needs at hand. The way I see it, FoFF's next generation work is among the most important things any farm advocate could be doing.

The average age of farmers in the U.S. is approaching sixty, and the training of the next generation of family farmers is way behind schedule. Mostly because there isn't a next generation of farm kids waiting to take over the family farms. With the shift to an industrial, chemical-based system of food and agriculture in this country, family farms became less and less viable economically, and industry simply buys them out and adds their land to the industrial machine. That's where FoFF comes in.

Nellie directs FoFF's iFarm and FarmON! programs. iFarm is a landlink program that connects farmers who have no land with landowners who are looking for a farmer to work with. Since its inception in 2009, iFarm has had over 700 landholder and landseeker listings and made over 50 land connections for farmers and ranchers around the state – a great track record for a landlink program. And FarmON!, coordinated by Beth Satterwhite, FoFF's Young and Beginning Farmer Organizer, is Oregon's chapter of the National Young Farmer Coalition, which strives to recruit and train beginning farmers. At present, both programs are still in a process of refinement – improving their systems and outreach. But they are in place and growing.

iFarm is the older of the two programs, and so far, iFarm's demographics have been a bit of a surprise to FoFF staffers. Their initial expectations were very much like my own. When I heard about the program, I assumed that old farmers nearing retirement would hook up with young farmers looking for a future. Hasn't necessarily gone that way.

"That's what we had expected most of the landholder

Like most of the Friends of Family Farmers staff members, Nellie McAdams is a farmer. She still works on her parents' hazelnut farm in Gaston, Oregon, helping her father do everything from equipment maintenance to nut harvest. This view, looking west from their front yard, shows the beauty of the rolling hills in the northern Willamette Valley. The orchards on the farm have been producing top quality nuts for decades.

listings to be, but it just hasn't happened," said Nellie. "In truth, it's been a real mix. A lot of landholders are not actually farmers themselves. Some of them purchase land specifically because they want to pass it on to young farmers. Some purchased land to live in the country and have that lifestyle, but they want someone else there farming. Others purchase land hoping to find partners to work with because they may not have all the needed skills. People are all over the map in terms of why they acquired land and what skills they have to offer. So mentorship doesn't necessarily come with the land contract, which is what we'd originally thought older farmers might provide."

Nellie believes some older farmers might find it difficult to fully participate because of a lack of comfort using a computer and doing everything online. I didn't mention it, but I suspect

there also could be a bit of a language barrier old farmers have to overcome. Because new young farmers just talk differently than old farmers. None of the farmers I knew growing up would have understood what it means to "be passionate about becoming a producer." I've heard two different farm interns make that comment during the writing of this book, and the second time it almost made me chuckle. Actually, the first time I had to ask what the person meant exactly.

I don't take the comment lightly, and I understand that they really care about the work they're doing. It's simply that most old farmers never have been involved with a "passion project." They simply lived on a farm and kept on doing what they grew up doing. Passion is a word very few old farmers use. So their comfort level with the iFarm program is most likely affected by a variety of influences.

But if anyone can figure all that out, it's probably Nellie. She's sharp. Energetic. Motivated. And her heart's in a good place – same with Beth Satterwhite – making them the perfect duo to spearhead FoFF's FarmON! Program.

As Oregon's branch of the National Young Farmers Coalition, FarmON!'s job is to host and promote events around the state that provide education to beginning (and potential) farmers, young and old. But Nellie is quick to point out that these events also need a social element.

 "The social component is so important," explained Nellie, "especially for resilience and preventing attrition, which happens quite a bit. A lot of these folks not only didn't come from a farming family, they weren't even rural. Most came from urban situations. Now, if they're living in a rural area, they're going through cultural changes as well as business changes. They're faced with building a whole new community. And while they may have a few friends in their area, they don't have time or energy to go out and have a beer, especially the first couple of years.

"FarmON! offers a way for beginning farmers to strengthen that community. And we're really not just talking about young farmers. That is the national organization's name, but we all sort of agree that's a little bit narrow. We are speaking directly to any aged farmers who feel they're beginners. And to do that effectively, we also need the experienced folks. Because this is a way for people to meet and share ideas and build the skill sets that beginning farmers really need."

The program's goals are to provide cheap, free, or at-cost skill shares that farmers in communities throughout Oregon have identified as a priority. Because input is coming from all over the state, the skills being shared will naturally have a little bit of a regional focus. Seed saving might be a topic in parts of the Willamette Valley or southern Oregon's Rogue Valley. Season extension could be a focus both on the coast and in eastern Oregon, which also would likely be interested in water conservation. Really, the topic doesn't matter as much as the demand for the topic, whether it be tractor mechanics or body mechanics.

"And of course there will be crop mobs," said Nellie, as though that was just a given and didn't actually need to be verbalized. "We're going to be getting people together on a farm and doing whatever that farm needs extra hands to do. Our job is to do the outreach and help get people there, and then there's a potluck afterwards so everyone can enjoy each other's company."

These on-farm get togethers will, in a way, lay the groundwork for what FoFF hopes will become an annual day-and-a-half celebration that brings together people from all over the state for skill shares, celebrations, good food and camaraderie.

By the time Nellie has finished sharing the vision of FarmON! and the Next Generation Program, I'm totally convinced that these events will be a central hub for Oregon's next generation of farmers and ranchers. Her energy is palpable. And her description offers glimpses into the way FoFF has learned to collaborate with other organizations to build critical mass and make things happen. Working closely with Rogue Farm Corps – one of the state's premier farm intern programs – helps assure its success.

Collaborations such as this are a central component of FoFF's overall strategy. It doesn't matter what farm or food organization I mention – Ten Rivers Food Web, Oregon Rural Action, Food Roots, OSU Small Farms Program – Nellie can name a way that FoFF has partnered with them.

"With OSU Small Farms, I'm on the Small Farms advisory board," said Nellie. "Plus, I'm coordinating five panels for the Small Farms School conference in September. FoFF helped coordinate with Ecotrust Small Farms and a number of regional organizations a series of day-long workshops with Atina Diffley about selling fresh produce to wholesalers. We always coordinate a panel at the big annual Small Farms Conference. We try to partner with most food and farm organizations in the state."

And FoFF also partners with individual farmers. Its annual Family Farmer and Rancher Day at the State Capitol, organized by FoFF Field Director, Leah Rodgers, is a perfect example of how that partnership can help effect change.

Family Farmer and Rancher Day is held in Salem, Oregon every other year during the legislative session. In 2013, FoFF

Friends of Family Farmers sponsors programs throughout the state that bring working farmers, would-be farmers, and consumers together to share information and advance the small farms agenda. Events range from panel presentations (like this InFARMation event) to full-fledged conferences that promote and celebrate sustainable farms.

brought about 150 farmers and ranchers from around the state to Salem to advocate for bills that help support sustainable food and farming.

"The morning was a briefing on different bills," said Nellie, "so people could learn about the details, ask questions, and decide which bills they wanted to advocate for with their particular legislators. We scheduled meetings with 80 out of 90 Oregon legislators, and we had constituents there to meet with them. We also set up a farmers market in the middle of the capitol building so anyone walking through would be exposed to the benefits of local and sustainable Oregon food, as well as giving people an opportunity to meet some of the farmers. Family Farmer and Rancher Day is a huge event for FoFF, and really, we believe our ability to get every one of our priority bills passed last year was the result of that grassroots effort."

One of the bills FoFF helped to get passed, under the leadership of Ivan Maluski, FoFF's Director and Legislative Director, banned commercial canola production in the Willamette Valley for five years, which was critical to help protect Oregon's vegetable seed producers. Another was the state's very first financing program for farmers and ranchers, called Aggie Bonds, which reduces their interest rate on qualifying loans. A third made it easier for small poultry producers to slaughter birds on their farms for sale directly to consumers.

In 2011, FoFF accomplished its first legislative victories, including enabling farmers to direct market their own minimally-hazardous value-added goods (like preserves and dried beans), and to process and sell on-farm up to 1000 poultry per year. FoFF is the only independent family farm organization working on these kinds of issues in Salem, and has been doing so since 2011 when FoFF carried the vision of their landmark guiding document, the farmer-created Agricultural Reclamation Act, to Salem for the first time.

The success FoFF is enjoying in its efforts to promote sustainable farming is impressive. That success has come from diligence and the ability to keep working even when progress is coming so slowly it's barely noticeable. Much like farming itself.

"One thing my dad told me about farming has always stuck with me," said Nellie. "He said a lot of people who want to farm are smarter than he is, but they don't all have the patience to do the same thing over and over and over again from sun up until sundown. And that makes all the difference." 🍴

(Clockwise from top left) FoFF merchandise. A crowd begins to gather at an InFARMation (& Beer!) event organized monthly in SE Portland by Leah Rodgers, FoFF Field Director and Deputy Director. A central program of FoFF's Eater Campaign, InFARMation brings the issues faced by Oregon's family farmers to the attention of urban consumers. Topics range from season extension to international trade. Nellie and FoFF Office Administrator, Erinn Criswell, welcome guests.

ABOUT

FRIENDS OF FAMILY FARMERS
Portland, Oregon

Friends of Family Farmers (FoFF) is a grassroots organization that promotes socially responsible, family-scale farming while fighting the growth of factory-scale corporate agriculture in Oregon through four major program areas:

Farmer Campaign, which is focused on farmer outreach, education, support and policy development. The Farmer campaign implements FoFF's landmark Agricultural Reclamation Act, a roadmap for sound family-scale agricultural policy created by and for Oregon farmers through dozens of listening sessions FoFF has held across the state since 2010;

Eater Campaign, which seeks to educate the public on the importance of socially responsible farm policies, and includes FoFF's flagship InFARMation (and Beer!) educational event, which has been educating eaters about the issues faced by Oregon's sustainable family farmers every month since 2009 in Portland, and the Pro Pasture campaign, which encourages consumers and retailers to patronize socially responsible ranchers who raise their animals outside on pasture;

Watchdog Campaign, which holds government agencies and policy makers accountable, ensuring that family-scale farmers have a voice when decisions that impact them are made. Upon request, this campaign also provides assistance to citizens groups that see their local agriculture and rural ways of life being threatened by industry;

Next Generation Campaign, which includes the online landlink program iFarm, and educational efforts focused on young and beginning farmers, including housing the local chapter of the National Young Farmers Coalition, called FarmON!.

For more information about FoFF programs, please visit their website at friendsoffamilyfarmers.org.

Vibrant Valley Farm

*Motivated by a desire to connect youth with food and farming,
Kara Gilbert and Elaine Walker are working to build their farm's finances
and support so they can introduce educational programming.*

Elaine Walker and Kara Gilbert met as freshmen at the University of Oregon. Both city kids – Elaine from San Francisco and Kara from Portland – they hit it off immediately and began a friendship that would enable them, years later, to reunite and begin building a new type of farm in rural Yamhill County, Oregon. The foundation for this new adventure was built through shared interests and experiences which began during those early days in Eugene.

During college, in addition to their urban roots, both young women shared a passion for helping people in need. They pursued interests in social justice, in creating opportunities within underserved communities, and in educating disadvantaged urban youth. They also both happened to enroll in a university course called Urban Farm, which taught them how to grow vegetables, exposed them to our country's ever worsening food justice issues, and allowed them to discover just how much they enjoyed working outdoors.

The Urban Farm instructors encouraged all their students to participate in the World Wide Opportunities on Organic Farms (WWOOF) program, in which people exchange labor for room and board on organic farms around the world, because doing so enables students to learn how organic farms in differing cultures and environments approach sustainable agriculture. Though not together, both Elaine and Kara stepped on this path.

While still an undergrad, Kara began her studies abroad in Italy, working on three different farms… "My goal was to work with three farm families and have them all be really different," she explained. "I was able to spend time learning from one couple who had a bit more experience, and then stayed with two younger couples who were just really going for it. Everything about that experience was cool, and after getting my undergrad degree I wanted to continue to travel, so I did the same thing in South America at a permaculture center in northern Patagonia."

After her time in Argentina, Kara spent time working at a special farm in Hawaii which offered a permaculture therapy program. Young people would go to the farm and work the land as way of creating change in their lives, or as Kara says, "It was basically a place to work out a lot of their problems, and it worked. That experience was inspiring for me."

After Hawaii, Kara returned to Portland to attend graduate school at Portland State University in a program called Leadership in Ecology, Culture, and Learning (now Leadership for Sustainability Education). And while she studied garden education in that program, she was running a sixth and seventh grade garden program at Lane Middle School in Portland, and continuing her youth therapy work. A three year stint at a working CSA farm following grad school brought her to the point of wanting to begin her own program on a working farm.

The farming experience they have gained working abroad and for other Oregon farmers is made apparent by the lushness of these mid-summer vegetables. With only the two owners doing all of the work, Vibrant Valley Farms maintains a vibrant CSA program and supplies a variety of restaurants surrounding their wine country farm.

Naturally, Kara called her best friend Elaine to see if Elaine wanted to take the leap with her, because she knew that Elaine had been following a similar track.

After majoring in journalism – "I thought it might help change the world if I could get better information out to people." – she found herself working in an office and not feeling good about it. So she left the office world behind and went to the permaculture center in Patagonia based on Kara's recommendation. While there, she was able to complete some projects that Kara had begun during her stay, and more importantly, she cemented her desire to work outdoors and learn to grow things.

"After WWOOFing, I went home to San Francisco and became a garden educator," said Elaine. "Initially I was working in Oakland with elementary schools and after school

programs. And then I spent a summer working at Pie Ranch in Pescadero."

Pie Ranch is a well known educational farm that works to bridge the urban/rural divide by bringing urban high school students to the country to learn about organic food and how to grow it. Pie Ranch also actively participates in the process of training a new generation of farmers who will work to integrate farms into the larger community and the web of nature.

"I soaked up as much knowledge and experience as I could at Pie Ranch," explained Elaine, "and then after spending some time back in San Francisco at a local garden, I enrolled in an agroecological school in Santa Cruz called the Center for Agroecology and Sustainable Food Systems. A lot of people who run CSAs, especially in California, have gone through that program. We ran a 75 member CSA, so it was very much a working farm, and it was great to get that hands-on experience. Then as I was finishing up that program, I knew I wanted to farm, but I wasn't sure where to go. That's when Kara called and said she had found some land in Oregon and that we should start a farm, and eventually, an educational program. So I moved up here, and we started Vibrant Valley Farm."

Now two years into their farm, both women admit it's been a challenge, but both also express complete happiness with their decision.

"It's hard, but I love this life," Elaine said. "You eat what you grow. It's very creative. It's outside. I'm my own boss. It's really nice to come up with our own schedule. We have so much fun working together. We take time off to make sure it's still fun, but we just laugh a lot. The best job I've had thus far, for sure."

"It is fun, but it's also important to be able to be okay with the lonely time," explained Kara. "I mean if you come from the city and you've got a lot of energy, you have to be able to step back and realize that the lonely time is a beautiful thing. And our culture has none of that… it's like, next screen, next screen, next screen… it's so intense. This helps you shed a lot of that. And another thing the farm teaches you is letting go. I'm getting into some philosophy over here, but it's true. It teaches you about your life. Your relationship with yourself. Your relationship with everyone else. Your relationship with the land, and that you cannot be attached to anything."

Elaine continues, "We always talk about the mandala of it. You work so hard to create this patchwork quilt of food… this artwork basically. And then it's turned into the ground. It doesn't disappear, of course, but you have to start all over. But each year is a fresh start, and you're ready because by Spring, you miss the smell of the dirt and that righteous tiredness that comes with all the Spring preparation. You're excited to get back at it. And that's why we take winters off, which is super important for us. I think people who do year-round farming are badasses, but for me, I know that would mean burnout if I didn't have winter off."

"Being able to make that choice is nice," adds Kara.

"I mean there are people all over the world who do this type of work and don't have a choice about taking time off. But I don't have to worry about making sure my village has broccoli. I just have to make sure a couple people have broccoli. And we're getting our farm systems perfected, so we don't have any problem getting that broccoli or whatever to our customers."

Both Kara and Elaine are clear about their goals, which remain the same from the day they started two years ago. Make the farm work first, then build in the educational component. Making the farm work means they had to begin as a CSA farm because it allowed them to begin farming even though they had extremely limited start-up capital. However, since launching their CSA, which they continue to grow, Vibrant Valley has begun acquiring additional customers from both the restaurant and grocery store ranks. Kara credits these gains to their outgoing personalities.

"We're good at going out and meeting people and pushing our products," explained Kara. "That's our strength. I mean we could sit here and feel sorry for ourselves because we're not selling enough, or we could say I'm tired and I want chocolate but instead I'm going to go hustle shishito peppers because we've got a ton of them. So we're actively out selling our product, and at the same time we're perfecting our systems, determining what works and what doesn't, and figuring out what's sustainable."

Both women agree that when they started, it was farm management they were least confident about. But the business gains they've made with their farm have brought them to a point where they will be comfortable as they begin to expand their vision.

"A big part of our original goal was to teach," said Elaine. "But we didn't want to start that program without a viable business. Bringing ten to twenty young people out here simply would not work if we don't know what we were doing. But now we're at the point where we can begin serious talks about how to add in education, so we're working on incorporating that element of it."

Step one will be figuring out what the community needs and how Vibrant Valley would play into that. And they will need to determine what demographic to work with. Being city kids, they feel especially connected to urban youth, so that's a distinct possibility. But their travels and experiences have made Elaine and Kara aware of food system injustices throughout society, so they anticipate looking at underserved populations everywhere, which could take them in a variety of different directions.

They anticipate that their search for educational partners will begin in Portland, and an area where Kara has extensive experience and relationships with a variety of schools and organizations.

A makeshift altar sitting outside the farm's cooler and staging area features a variety of flowers and found objects, as well as a happy Buddha. The spiritual qualities of their farming lives are as important to both Elaine and Kara as the happiness they gain from working outdoors.

"Once we determine who our allies are and who we can work with, then we figure out how our farm can fit with therapy or job training or whatever we decide to focus on," explained Elaine. "Because based on our experience, we've seen how a farm can be everything. Beyond growing plants, farming teaches marketing, accounting, even floral design or event planning. There's so much that fits in a farm, and as educators, we use this as our stage. But the final program

will depend on the age and background of the population we're working with."

Lofty plans. But will they work? The combination of energy, pragmatism, and passion are difficult to bet against.

"It's important to not get caught up in the idealistic young farmer mentality of just needing to do something and not worrying about making money," said Kara. "You can definitely make money, and, in fact, you have to in order to keep it going. It's just a matter of figuring out what that looks like. And part of that is letting go of what doesn't work and embracing what does. And we can't just run ourselves into the ground. That's the martyrdom thing you have to avoid. In order to farm or teach well, you have to stay fresh. That's the only way it will work."

Elaine sums the conversation up with a look into the future. "I think I will always work with food, in one way or another. Food is everywhere. It could be on this land, on this farm. Or it could be elsewhere. I simply can't guarantee where. But for me personally, I want to work with young people and food, no matter what."

(Clockwise from top left) Statice blossoms for use in floral designs. A look back across the field to the farm's office and staging area. Snapdragons burst with color.

ABOUT

VIBRANT VALLEY FARM
Elaine Walker & Kara Gilbert
McMinnville, Oregon

Vibrant Valley Farm was founded by Kara Gilbert and Elaine Walker in 2012 after both women had spent several years working as farm laborers and agricultural educators. Their goal for Vibrant Valley is to establish it as a working CSA farm, then gradually partner with local schools and youth projects to create mentorship programs as well as green job training opportunities.

Unlike many CSA operations which essentially focus on growing food, Elaine and Kara strive to connect the process of growing food with the greater community. With backgrounds in placed-based alternative education, they are working to create an inclusive environment on their farm where young people will be able to share in the agricultural process and gain hands-on learning experiences.

The farm currently offers vegetable CSA shares with multiple pick-up sites in Southeast Portland, as well as providing produce to area restaurant accounts. Elaine and Kara also are beginning to expand the number of flower varieties grown on the farm with an eye toward expanding that market.

For more information about Vibrant Valley Farm, visit their website at vibrantvalleyfarm.com.

Adelante Mujeres Sustainable Agriculture

The Sustainable Agriculture program at Adelante Mujeres is an educational program that provides the Latino community of Western Washington County with increased access to healthy food.

When I first met Alejandro Tecum, we established a point of connection by sharing that we both grew up on family farms. Mine in Kansas, and his in Guatemala. We had something else in common, as well. We both grew tired of farming when we were young and wanted to do something else. Then later in life we rediscovered our passion for agriculture. But he tells his story in a far more entertaining way.

"I farmed with my father, but I didn't really like it," Alejandro said. "Every year we had to dig the soil. Year after year. So I got tired of that, and I kind of left farming. But when I got married, my wife said to me, 'We're married now, Alejandro. It's time to farm.' So I started farming again. Then I lost my first wife, but after a while I married again, and my second wife said, "It's Spring, Alejandro. It's time to dig the soil.' So I kept farming."

Alejandro carries humility and his excellent sense of humor with him as he directs the Adelante Mujeres Sustainable Agriculture program from the organization's Forest Grove, Oregon offices. He also brings a quiet passion to his work that is helping to shape a new wave of sustainable farmers within Oregon's Latino community. It appears that his history with farming — a gradual transition from pre-Columbian to conventional to sustainable methods — enables him to speak effectively to all members of the Latino community about the importance of food and how it is grown.

Alejandro's evolution began at home in Guatemala. His father owned the equivalent of about two acres of land, and the family turned the soil by hand every year. When he was very young, his family planted in what he calls the "pre-Colombian way of sowing. We planted corn, beans, fava, and squash all together. It was not easy, but it worked for us."

But things changed about the time Alejandro was becoming a teenager. "A Peace Corps volunteer came to our place, and he taught that there was another way to grow food faster. He brought chemicals. He said you have to use this fertilizer, and pesticides, and fungicides, and he gave my father these chemicals and seed potatoes to plant. And that year, we had so many potatoes. People got excited, and that's how our people got started using chemicals. But the next year, we had to use more fertilizer and more pesticides and fungicides to get the same result, and that pattern continued.

"I understand now that they are killing the microorganisms in the soil. The soil is dying. And still today, they are using those chemicals. I would like to go back down there and tell them that they are doing it the wrong way. Because now I'm convinced that using sustainable techniques is the best way. Once we understand how the nutrients get into the plants and how important the life in the soil is, it's easy to see that farming with chemicals or using GMOs is not a good way to do it."

Alejandro learned sustainable agricultural methods after

Graduates of Alejandro's Sustainable Agriculture program are successfully selling direct to consumers at multiple venues. In the photo at left, Norma and Nicolas Amaro of N&N Amaro Produce launched their farm business in 2006 and now sell at markets throughout the Portland metro. At right, Sabino Amaro, of Amaro's Produce, is ready to greet customers at the Forest Grove Farmers Market. Sabino and his wife, Reyna Rojas, launched their farm business in 2005, the first year of Adelante Mujeres Sustainable Agriculture program.

emigrating to Cornelius, Oregon, from Guatemala. One of his first teachers was Will Newman from the Oregon Sustainable Agricultural Land Trust (OSALT). And in addition to classroom and hands-on training, he has spent a tremendous amount of time researching sustainable methodologies online. He mentions the writings and videos of Dr. Elaine Ingham, chief scientist at the Rodale Institute, as one of his favorite resources.

But after so many years of practicing agriculture using different methods, he admits he was skeptical about what he was hearing. So he established test plots in his home garden to test the information.

"I am a doubtful person," admits Alejandro. "If I don't see it, I don't believe it. So I created a space to test the sustainable techniques I was learning, but kept on with my conventional methods on the rest of it. But after a year, I saw the difference. So I continued the same way the second year, and the difference in my two plots was very clear. After that, I was convinced, and now I keep learning as much as I can about how and why the process works."

Understanding the need to see proof helps Alejandro be patient when he introduces the ideas behind sustainable agriculture to his students, many of whom have no experience whatsoever with farming or gardening. Alejandro believes that not having experience growing food can actually be an advantage in some way. In fact, he says the students with no experience usually are more receptive to his messages. But experienced or not, all his students enjoy his sense of humor and the many entertaining examples Alejandro uses to help illustrate the concepts behind his techniques.

"I have a bad joke about mothers-in-law I tell when I teach the weed management class," he said. "Of course, I don't mean my mother-in-law, who I like very much. But I tell my students that weeds are like mothers-in-law. They are always there. So what we have to do is learn how to deal with them and live in peace. And people laugh, but they get the idea. And then when I have their attention, I tell them that when I walk through my own garden, I deal with the weeds by pulling them and tossing them on the beds. That works especially well in summer. If you pull a plant on a sunny day, by the end of the day it is dead, and then it turns into soil."

Experienced farm workers are usually less receptive when they first begin attending Adelante Mujeres Sustainable Agriculture classes. These are people who, like Alejandro, have grown up with conventional chemical agriculture and have so many doubts to overcome that it holds them back at first. But he says the ones who give sustainable methods a fair test are convinced in time.

"Most of the experienced farm workers who stay with the class want to start with a small plot in the community garden," Alejandro explained. "They need to test the techniques because they don't believe it. But after they see the benefits, they want to grow more and some would like to start their own farm. Of course, then there is a new challenge, which is finding land. They don't have the capacity, the money, to buy their own land, so that's a problem. There are landowners who are willing to make some land available, but most of them are only willing to commit to one year. And that is not enough time."

Alejandro explains that it takes at least three years to effectively convert a conventionally farmed field to a sustainably farmed field because of the amount of damage that must be undone. The process of building new, healthy soil that is alive and filled with microorganisms is a natural process that mimics nature. Year after year, nature incorporates new organic matter and the soil is gradually renewed. To truly

benefit from that renewal, a sustainable farmer would need to farm the land at least five years.

Even if land does become available, the lack of adequate language skills is another challenge for many immigrants who wish to farm. To help alleviate that challenge, Alejandro has worked with Nellie McAdams, the staff attorney for Friends of Family Farmers (FoFF), to develop a standard contract that students can use to enter into lease terms. Adelante Mujeres also works with FoFF to host presentations where potential landlords can hear from other land owners who have had positive experiences leasing land to beginning farmers. And both organizations hope to soon provide sessions to help educate beginning farmers about how to successfully navigate the process of leasing land.

For those farmers who do begin producing food, the Sustainable Agriculture program operates both a CSA for community members and a wholesale produce distribution system that provides sustainably grown produce to local institutions, restaurants, schools, and local businesses. Plus, as manager of the downtown Forest Grove Farmers Market, Adelante Mujeres can provide that additional market outlet for the program's farmers. Several of Alejandro's former students sell vegetables or operate as hot food vendors at the market.

Alejandro has high hopes for the farmers who want to grow a business, but he knows that the majority of his students will only farm on a very small scale. Primarily in their own yards. But that's okay because he believes the most important outcome of his work is that everyone, regardless of their situation, will have access to good quality, healthy food. He understands that to produce that food, the first step is healthy soil. But it's the food, itself, that he uses to get his students' attention.

"In my classes, I share healthy, sustainably grown food with the students, and they recognize the difference between that and the chemically produced food in supermarkets right away," he said. "Then they start to get interested in how that food gets produced. It is then that I can begin teaching them about how to build healthy soil. But it starts with the food, because we all want to eat good food."

Then almost as an afterthought, Alejandro stresses that there is another critical element required to successfully teach sustainable farming. One that is not easy to hide if it is missing.

"One of the most important things about teaching these methods is that I, myself, am convinced about this being the right way to farm. If you are not convinced, you can't teach it. I was talking recently with a woman who wants to run a similar program in another city. But as we talked, I sensed that she was not completely convinced about all the sustainable methods. You cannot teach it if you do not believe it and have practice doing it."

The Forest Grove Farmers Market opens each Wednesday afternoon in downtown Forest Grove, Oregon. Since 2005, Adelante Mujeres has managed the Forest Grove Farmers Market, fostering cross-cultural exchange during this family-focused community event. The market sets up right outside Adelante Mujeres offices on Main Street. In addition to the farmers market, the Sustainable Agriculture program provides its students with the opportunity to produce food for the organization's CSA and wholesale food distribution service.

is growing, in the size of the classes, but also in giving away knowledge to people. I think we are helping people understand that farming is a valuable profession. It is not something to be ashamed about. People in society often look down on farmers and believe that they are not as good as people in other professions, but that is not true. Farming is very important work to do."

As a teacher and farmer, Alejandro presents a convincing case for sustainable agriculture, something he almost apologetically confesses he has developed a passion for… both doing it and teaching it. In a way, he believes it is fate that has brought him to this point in his life. But regardless of how and why he got here, he is very happy to be here.

Replicating Adelante Mujeres Sustainable Agriculture program in other communities is a topic of some interest within the sustainable agriculture community. Alejandro mentions several communities in Northwest Oregon who are considering similar programs. Plus, he confirms that he is frequently interviewed or asked to speak about the program. He believes in his program, so it pleases him that it is attracting positive attention. But he makes it very clear that his principal concern is continuing to strengthen the programs of Adelante Mujeres.

"I am here because Adelante Mujeres pays me," said Alejandro. "And I am very grateful for that. I have come to love this work very much. We are happy that the program

"It is interesting to me that my desire after high school was to become what I think you would call an agronomist… a soil scientist. But my family couldn't afford to pay for that school, so I became a teacher instead. I'm thinking now that it was a good thing that I didn't go to that agronomy school, because I would have learned the conventional agriculture methods. Now, all these years later, things have come back to learning about soil. So in some ways I am doing now what I wanted to do when I was young, but this time it is sustainable methods. Yes, I like this work very much, and I hope I will be able to keep doing it for a long time."

(Clockwise from top left) Chili peppers ripen in the autumn sun. The Forest Grove Community Garden provides space for some Adelante Mujeres program participants. Vegetables that comprise a portion of N&N Amaro Produce's market display.

ABOUT

ADELANTE MUJERES SUSTAINABLE AGRICULTURE
Alejandro Tecum
Forest Grove, Oregon

Sustainable Agriculture is a program that strives for social justice and equity through ecological land management and economic vitality. It is one of a variety of programs created by Adelante Mujeres, a Washington County, Oregon non-profit that provides holistic education and empowerment opportunities to low income Latina women and their families.

The Adelante Mujeres Sustainable Agriculture program provides aspiring Latino immigrant farmers and gardeners with the training and skills necessary to grow produce using sustainable methods and to successfully market their products.

Participants enroll in a 12-week sustainable farming class, taught in Spanish, that covers topics like farming techniques, soil maintenance, crop planning, and pest management. The course also includes practical workshops and opportunities to participate in farming related conferences and networking events.

You can learn more about Adelante Mujeres on their website: adelantemujeres.org/agriculture.

Abundant Fields Farm

Abundant Fields Farm is receiving the support of a business incubator process in much the same way other types of start-ups do. Sharing infrastructure with other beginners helps make it possible.

The recently developed incubator program at Headwaters Farm gives new farmers a four-year period to kick off their new farming career. The objective is to use these four years as a springboard to help them advance their agricultural business model. Rick Reddaway, a member of the incubator's inaugural class of 2013, is nearing the completion of his second year in the program. And according to Rick, the first two years flew by pretty fast. It is now occurring to him that his final two years will probably speed by as well.

He feels fortunate to have two years left to continue building his markets and to identify a property where he can keep his nascent operation growing. Change has been a constant for Rick and his wife, Heather, over the past several years. As such, this latest seismic shift should be something they can weather.

A little more than three years ago, Rick was working as a project manager for a manufacturing company. But even though his career was on the rise, his happiness wasn't. He and Heather longed to return to a life outside the city, enjoying country life and gardening or farming... something similar to the way they both had grown up.

For Rick that childhood took place on fourteen acres in West Linn, Oregon. To his dad, it was a hobby farm, but to Rick, it was just a farm... the place where he first learned to garden and tend animals. Heather grew up in Sandy, not far to the east of where they live now at Headwaters. Her parents gardened on three acres there and still do. Most likely, for both Heather and Rick it was the lifestyle they remembered best. Life in the country "digging in the dirt" is how Rick describes it. They wanted to give their new son a childhood like the one they both had, and every farmers market they walked through honed that desire more poignantly.

"We didn't want to continue to live in the city, and I had reached the point where I just couldn't sit at a desk anymore," said Rick. "So we talked about it and decided to make the leap. We moved in with Heather's parents and they set aside a quarter-acre plot for me to get started growing vegetables."

Rick didn't waste any time getting started. As soon as they moved he started planting, and he applied to several farmers markets. The Montavilla Farmers Market said yes, providing him with a sales venue. Fortunately, Heather's mother was willing to act as his mentor, imparting her decades worth of experience and helping him develop a planting plan... the first of many new skills he would discover he was lacking.

"I had gardened before, obviously, but that mostly consisted of planting some seeds and growing enough food to satisfy myself," explained Rick. "Planting enough to fill an entire season at a farmers market and growing produce that will satisfy shoppers was a whole new thing. In hindsight, I'm sure I could have benefited from doing an internship or

Looking across Rick Reddaway's vegetables to his greenhouse, we see the majority of the farm's one-acre plot. Rick's construction ability – he built his own greenhouse – helped him land the role of incubator farm caretaker in addition to one of the farming opportunities. The incubator, which is called Headwaters Farm, is operated by the East Multnomah Soil and Water Conservation District. The environmental and conservation emphasis of EMSWCD leads it to place a heavy emphasis on organic methodologies and land stewardship in the training programs their incubator farmers participate in.

working on a farm for a year or two. In fact, I'd probably recommend that to anyone else just getting started, but I've always been the type to just jump in and do it. There's some pain involved, but it's good pain."

It was at the Montavilla market where Rick met fellow farmer Rowan Steele, who helps his wife, Katie Coppoletta, run Fiddlehead Farm. Rowan also manages the Headwaters Farm incubator program. Rowan saw enough promise in the work Rick was doing to suggest that he apply to join the first set of incubator farmers at Headwaters.

Rick saw an incubator program as a good opportunity, but to transition from the quarter-acre plot at his in-law's house to a formal business incubator meant he would have to stop shooting from the hip and start getting organized. Step

one was writing a legitimate business plan.

"I had never done a business plan," Rick confessed. "But now I had to because I was applying to start a business. I mean there was some crossover with my project management work. There are processes that have to happen, but now I had to apply that thinking to farming. In the beginning, I guess I was too caught up in the dream to think that way. I just thought I would go start planting a bunch of stuff and sell it. But as the idea of farming started to become the reality of farming, I was encountering logistics that hadn't occurred to me."

After his application and business plan made the cut, the next step in the incubator process was the interview. "They were throwing questions at me, and that reinforced the seriousness of it, but I felt okay about the interview after it was over. And then I got the invitation to join the program. Man, I was ecstatic."

That was how Rick's incubator experience began. To sweeten the pot, he also was chosen to live onsite at Headwaters and act as the property caretaker. He actually was the only program farmer to apply for that role, but regardless, he sees it as a plus. He, Heather and their young son, Brenner, live on the sixty acres with low rent a trade for work that ranges from helping with construction projects to making a daily walk around the property to check on things.

"My boy loves those daily walks," Rick said. "He calls it a walk down woolly bear lane because we're always seeing those little woolly bear caterpillars everywhere. We wanted him to have a chance to grow up in the country, and right now he has a sixty acre backyard."

At least for two more years anyway. What happens after that remains the question that simply won't go away. The answer will probably involve money, which Rick admits in spite of the fact that he's steadfast about the fact he's not traveling this path for the money. "What I'm doing now directly touches other people. That matters. And digging in the dirt makes me happy. But I understand we have to make enough to keep it going."

In farming, as in most business endeavors, increasing revenue is the result of successfully expanding your market. But when one is just beginning, it's hard to know how much expansion is needed. Is the goal to find an acre to lease somewhere? Or maybe shoot for the moon and actually try to buy a dozen acres with a house. Most likely the reality lies somewhere in between the two, but only research will provide an accurate gauge. So to get a sense of what kind of money might be needed and what opportunities might realistically exist, Rick has begun exploring the possibilities.

Recently he spoke with a representative from the United States Department of Agriculture's Farm Service Agency about one of that agency's farm ownership loan programs. What he discovered is that he's a long ways from where a loan manager would want him to be in order to achieve the high end of the success scale.

"He asked me what my gross sales were," related Rick, "so I told him where I was. He didn't say anything right then, but a little later he said he'd like to see our gross sales in the $30,000 to $50,000 range to start talking about a farm loan. I'm nowhere near that, and I'll admit that kind of put a damper on my dreams a little. But I have to keep in mind that's just one loan program. There are lots of other possibilities."

Those possibilities include operating loans and micro loans among other things. Rick believes getting a micro loan to help with operating expenses might be a good place to start. That would enable him to develop a successful credit relationship and set the stage for something larger like a farm down payment loan down the road. An alternative approach is to step back from the 'perfect world scenario' and realize there are a lot of ways to be happy farming.

Rick related the story of a young farmer he knows who went through Oregon State's Beginning Urban Farmer Apprenticeship (BUFA) program but didn't have a solid plan when he finished. The guy just started looking around and found a small produce market on the east side of Portland that had three-quarters of an acre of land sitting untouched behind it. So he reached an agreement with the market owner to farm that three-quarters of an acre and sell his produce at the market.

Then there was another incubator farmer, a couple actually, who contacted the owner of a vacant lot right on Portland's Stark Street and asked if they could farm it, and the owner said okay. So now they farm a lot on Stark and an acre at Headwaters.

Telling these stories has to help Rick feel a little better about the future. There was nothing special about these other farmers, so there's no reason for him to believe things won't work the same for him. But he also acknowledges that one of the most important keys to any opportunities he may encounter is more sales, which means scaling up his production and improving his marketing efforts.

Currently Rick sells his produce at two farmers markets, and there are a lot of positives in that. He points out that he gets retail prices, and the time he spends at the markets interacting with his customers is one of the most fun and fulfilling things he does. Unfortunately, there's a limit to how much he's been able to sell. So he's starting to talk to other farmers to glean ideas and learn about other directions he might be able to go.

"Restaurants are something that I'd like to get into," said Rick. "One friend of mine sells to restaurants exclusively, and he's doing great, so I'm trying to learn what I can from him. I'm also looking at a few core crops I might be able to specialize in a little more, like peppers. I've recently started growing peppers for a local hot sauce vendor called Marshall's Haute Sauce Company. It's just one small company, but a hundred pounds of serrano peppers is a lot for me. Plus, I just started talking to another small hot sauce maker who's a vendor at the Montavilla Farmers Market. His business is called Hard Times Hot Sauce, and we've begun making some plans for the next growing season."

Rick Reddaway's background in graphic design led to a stylish logo for his new farm. And he hopes that a horseshoe he uncovered on the property brings him luck. Actually, all of the farmers at Headwaters should feel lucky that they were looking for a farm at the same time the incubator was starting up.

In addition to expanding the production of some of his core crops, Rick is considering diversifying by possibly adding meat birds or some type of animal component to his repertoire. Headwaters does allow its farmers to include livestock in their business plan, so this seems like it would be a negotiable possibility for him. On the other hand, both expanding crop production and adding other dimensions to his operation begs the question of just how much can one person do? Which means that adding an employee or seeking some type of partnership then becomes one more challenge that Rick knows must be faced.

"There's all this stuff that I never thought about when I started this," he said. "I know I keep coming back to that, but it seems like there's something new all the time. But, you know, every discovery I make, I'm still excited about it. I feel like every step is a measure of success for me. It is kind of scary at times, but it's all a positive. Out of everything I could be doing, I prefer this. And I do believe that we'll have our farm, one way or another."

(Clockwise from top left) Kale, which is a Pacific Northwest mainstay. Rick's lettuce mix is one of his best selling products. A beautiful red cabbage.

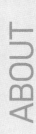

ABOUT

ABUNDANT FIELDS FARM
Rick Reddaway
Gresham, Oregon

Abundant Fields Farm is a part of the East Multnomah Soil and Water Conservation District's farm incubator program, which is called Headwaters Farm. Headwaters is located in eastern Gresham, Oregon, a suburb of Portland.

Rick Reddaway farms a one-acre plot at the incubator location, growing mixed vegetables using fully organic methods. Because the farm is not available for organic certification, in 2014 he applied for Certified Naturally Grown recognition, and following inspection, was granted that certification.

Rick's farm provides a broad mix of fresh produce, from kale to carrots to peppers to garlic, all of which are available at the Moreland Farmers Market and the Montavilla Farmers Market, both located in southeast Portland neighborhoods.

For more information about Abundant Fields Farm, please visit the website at abundantfieldsfarm.com.

Naomi's Organic Farm Supply

Walking into Naomi's Organic Farm Supply is like walking into a fun house. The colors, the energy, the happiness, and the commitment to a healthy, meaningful life is a big part of what this shop is all about.

Naomi Montacre was born thirty-seven years ago in Milwaukee, Wisconsin, where she was raised by a single mother. Not on a farm, but, like so many of the people who now are involved in sustainable agriculture, she and her mother did have a garden. So she had a sense of what real food looked like.

When she was young, Naomi thought she would grow up to be something really serious, like a lawyer or a criminal psychologist. Then she took an art class, and things changed. She started making art, got on a creative path, and ended up in the film industry doing prop master work in New York. But living in that world offered challenges Naomi wasn't interested in pursuing, so in a round-about way, she began a search for something she felt she could commit to. Something that was in a place where there weren't quite so many people.

She made a list of what she loved. Being outdoors. Being physical. Living a healthy, meaningful life. And food. Definitely food.

Almost everyday she would ride her bicycle past an urban farm that used to be a basketball court, and she found herself formulating a plan. She didn't care about making a lot of money, but she did want to live in a beautiful place with progressive-thinking people. Gradually, that urban farm began to grow in her thinking. And Naomi's planning process eventually brought her to Oregon and led her to attend Oregon State University's annual small farms conference. She had decided that farming made sense for her.

She ended up meeting a farmer who agreed to lease her and a partner some land, and she jumped into the farm movement wholeheartedly and began farming. She was working for herself, working for the farmer who leased land to her, and working for a third farm, as well. It was challenging, but she felt things were going well and she was on the right track. Then she broke her foot in a farming accident and her new life began to quickly unravel.

"I ended up moving to Portland and putting the farming on hold," said Naomi. "It was frustrating, and I lost my farm, and when I was recuperating I found myself thinking... 'if only I could buy a house and have a big garden I'd be happy.' But finding land to grow on was so difficult. I mean it was hard in the country, but in the city, everything was so expensive. But one thing I learned from that experience is that when you start down a path and something goes wrong – like for me when I broke my foot and couldn't farm I thought it was horrible – sometimes something really good can come from that."

The goodness that came to Naomi took many forms, but fundamentally, the series of events that followed the loss of her farm led her to realize that she could remain attached to food and farming without necessarily being a farmer. They also led her to a better understanding of who she is and one of the core principals that informs much of what she does and what she hopes to do. She basically really likes people.

"When I was living in Brooklyn, I pretty much forgot about that," explains Naomi. "It was just so crowded, and it was all too much. I convinced myself that I just wanted to get away and live in solitude. That took me to the farm, which I

In locating her farm supply company, Naomi Montacre wanted to be close enough to surrounding neighborhoods that home gardeners could easily access it, while keeping within the inner industrial area that affords her more freedom in how she handles shipping and receiving and also encourages regional farmers to stop by when they're in town on other business because she's centrally located. Good customer relations combined with Naomi's strong relationships in the farming community – she used to be a partner in 2 Donkeys Farm – has helped her develop a devoted following.

loved. But after moving back to Portland, which ironically I was kind of afraid to do even though it was a much smaller metro, I got involved with people in a different way and realized that helping people and working with people is just so much a part of who I am. And the fact that this is a slower, smaller, loving city has something to do with that, too."

Another thing that happened was that Naomi was encouraged to start a shop of her own around her love of organic agriculture and food production. Her mother moved to Portland to join her, and together they came up with the first version of Naomi's Organic Farm Supply.

"Over the years things have evolved, but the core principal has always been the love of people and what they're doing," she said. "The capitalist part is harder for me. It works

out, thankfully. I have fantastic employees. The demand and the supply is there. And it's beautiful. But if it wasn't for the people, I definitely wouldn't want to be involved in retail. Cultivating relationships and watching people and their gardens and farms grow. Watching people's children grow up. All those parts are really beautiful. Instead of four seasons, you get to see the seasons of people's lives."

Naomi's customers range from people who have created a little raised bed and are testing gardening for the first time, to sophisticated urban farmers who farm leased or rented land on larger plots in the city, to urban agricultural collectives, to organic farms in the surrounding countryside. Regardless of the size of the operation, Naomi admires the commitment gardeners or farmers of any size are making.

To explain that admiration, she gave me an example of what it takes to start a bike shop as opposed to an urban farm. A bike shop owner might have a five year lease in a fixed space. It's a climate controlled space, and there are a fixed number of variables. So any bike shop that can identify and address those variables probably has a better chance of surviving. Whereas someone starting a farm could have greater lease insecurity, water issues, or someone spraying pesticide next door. They face the process of improving the soil, dealing with varying pests, and the weather is a constant, unpredictable factor. So many factors are outside of the farmer's control.

"Going into a knowingly low wage enterprise with this crazy amount of variables," said Naomi, "it seems natural that a lot of people or groups of people would choose to or be forced to give it up at some point. For someone to even give it a try is amazing. But for those who do it for a long period of time, it's takes a genuine commitment and it's really a letting go process. Just being able to roll forward with things in spite of all the challenges and variables."

For Naomi, the learning curve didn't involve adjusting to the weather as much as it did adjusting to the vagaries of real estate zoning and development. Her first shop was located on a large parcel of land in outer Southeast Portland that had her dreaming of demonstration gardens and relationships with all types of organizations. But when the landowner preferred to see a different type of development on that property, Naomi lost her lease and had to search for a new location.

She ended up in a centralized, industrial area that meets her needs in terms of being able to have large trucks park in the street and unload cargo, but severely restricts her ability to sell plants or other products that need outdoor display space.

"There was a pretty major trade-off when we located here," she shared, "but overall it's worked out fairly well. We've found that there's an advantage to being more centrally located. A lot of our farm customers come to the city for a variety of reasons. Partly to do something related to their business, but also to have fun or to visit the dentist. They come from the coast, or Washington, or the southern valley. We even have a few customers from eastern Oregon. Being in the center of the city makes it easier for them to do what they came here for and also to come see us. So everyone's happy."

And keeping people happy is important. Naomi says that most happy farmers she knows share her ability to be flexible. They also tend to have very positive relationships with their customers, whether they are CSA members, or a dedicated groups of chefs, or simply farmers market regulars. Even if it's on the wholesale side and not directly with the eater, relationships are important.

"It's like that for us here, too," she said. "Without the relationships, all we would have here is a bunch of stuff. That just seems weird… selling a bunch of stuff just to be selling it. It's the people we work with… our relationships with these people who are doing amazing things. That's why we're here doing this work."

In addition to selling products, a significant role Naomi and her staff fill is that of an information clearinghouse.

"It's great hearing what works for people and being able to share that information with others," Naomi said. "Or what doesn't work. Like hairy vetch works as a cover crop for some farms, but doesn't work for a vineyard because it can crawl up a trellis. Every time we talk to someone we're learning about different experiments people are doing, and we know who they are and what their conditions are. We work with plenty of certified organic growers… including perennial systems like vineyards. But we also work with smaller scale farmers who don't have the sales volume to justify certification, but who are just as rigorous and need the same types of information. Over time, farms figure out what processes work best for them, and what size and shape they need to be."

In addition to supplying and networking with small and large scale farms, Naomi's Organic Farm Supply also provides a valuable information service to homeowners looking to get more in touch with their food, or even to try out a new food trend. Like raising chickens. Naomi says chicken interest was at a fever pitch in the not too distant past, but fortunately that surge of interest has leveled off as people have become more level headed about it.

One factor that cooled the chicken trend is the fact that people discovered chickens live a long time. Often well past the age at which they stop laying eggs. They're like having a cat or a dog. Many people also had to learn that eggs are seasonal. Laying eggs is a daylight sensitive phenomenon, and as daylight diminishes, chickens lay fewer eggs, often not laying at all through the winter months. Gaining this knowledge is important for people who want to eat locally and eat seasonally, but it surprises a lot of folks. For this reason, the Farm Supply offers workshops on a variety of topics, including chickens.

An organic seed display featuring Uprising Seeds, a company based in Bellingham, Washington, and southern Oregon's Siskiyou Seeds. Photo courtesy of Naomi's Organic Farm Supply

Her efforts to share knowledge and build relationships has helped Naomi get past the disappointment of not farming. She works with both plants and animals. Spends a great deal of time outside. Stays active loading bags and bales for her customers. She has come to feel much more content over time. And she gardens extensively at home, maintaining a personal relationship with plants rather than a commercial one. But is maintaining this retail operation something she feels she can continue on a long-term basis?

Everything appears set, at least for a while. So can we expect to see any significant changes at Naomi's Organic Farm Supply anytime soon?

"We're looking at that," she said. "We're kind of at a juncture right now where, like a lot of farms, where we don't know whether to grow or to shrink or try to stay the same size. We don't own this building, and we're only in a portion of it right now. So do we try to expand to fill more of it? If we do, that means more merchandize and adding more employees. What happens then if we have to contract later? One of my most important goals is providing all of my co-workers the best quality of life possible. Maybe that means special scheduling, or three days off a week, or whatever. But it especially means as much security as I can give them. These are people who have done other serious things and are choosing to be here because we all believe in the same things. So whatever we do, it has to work for everyone."

It's clear that Naomi's concern for the welfare of her staff is sincere. She talks about the positive energy they all try to develop among themselves so they can carry that to their customers. And the winter planning sessions when things are less busy and improvement ideas get added to a permanent white board in the office. The shared values they all have related to money… making enough to keep things going but not actively pursuing more than they need. So the key is not to be growing the operation financially, but rather to make sure that everyone is happy.

"We try to bring a non-profit attitude to this. To maintain that moral core, and ask the questions… Do the customers feel happy and well served? Do the people working here feel happy and not feel like they're being pressured in any way? Those are the pieces that are most important. Otherwise it's just not worth it."

(Clockwise from top left) Organic soil inputs in home garden package sizes. An energizing tool display. Organic garlic ready for Fall planting.

ABOUT

NAOMI'S ORGANIC FARM SUPPLY
Naomi Montacre
Portland, Oregon

Naomi's Organic Farm Supply was founded by Naomi Montacre, and is located in an industrial neighborhood in inner Southeast Portland.

According to its website, it provides organic soil amendments, fertilizers, compost, potting soils, animal feed, bedding, seeds, tools and books to supply both urban and rural farmers, gardeners, and landscapers. It also carries multiple organic feed options and one non-organic layer feed, from both larger and smaller Oregon and Washington mills. And it is committed to carrying only non-GMO cover crops, specializing in organic, along with organic herb, flower and vegetable seeds, the majority of which are grown in the Pacific Northwest region.

It is a goal of Naomi's to serve as a hub of products, information, and networking for those interested in organic gardening and farming.

For more information about Naomi's Organic Farm Supply, please visit the website at naomisorganic.blogspot.com.

ADVOCACY AND EDUCATION

Adelante Mujeres   adelantemujeres.org

The Cornucopia Institute   cornucopia.org

Friends of Family Farmers   friendsoffamilyfarmers.org

Oregon State University Small Farms Program
smallfarms.oregonstate.edu

Organic Consumers Association   organicconsumers.org

Women Food & Agriculture Network   wfan.org

CERTIFICATION

Certified Naturally Grown   naturallygrown.org

Oregon Tilth   tilth.org

Sustainable Farm Certification International
sustainablefarmcert.com

FARM INTERNSHIPS

Rogue Farm Corps   roguefarmcorps.org

World Wide Opportunities on Organic Farms
wwoof.net

FINDING LAND

Farmland Information Center   farmlandinfo.org

iFarm Oregon   friendsoffamilyfarmers.org/?page_id=130

International Farm Transition Network
farmtransition.org

GENERAL FARMING INFO

NAL Alternative Farming Systems Information Center
nal.usda.gov/afsic/pubs/ofp/ofp.shtml

National Sustainable Agriculture Information Service
attra.ncat.org

Oregon Department of Agriculture -
National Organic Program
oregon.gov/oda/programs/MarketAccess/MACertification/Pages/
NationalOrganicProgram.aspx

Organic Farming Research Foundation   ofrf.org

USDA Natural Resources Conservation Services -
Organic Farming
nrcs.usda.gov/wps/portal/nrcs/main/national/landuse/crops/organic

MARKETING AND DISTRIBUTION

Food-Hub   food-hub.org

PUBLIC DOMAIN SEEDS

Open Source Seed Initiative   opensourceseedinitiative.org

Organic Seed Alliance   seedalliance.org

Organic Seed Growers and Trade Association   osgata.org

SUSTAINABLE FOOD

Cooking Up a Story (videos)
youtube.com/user/cookingupastory

Ecotrust   ecotrust.org/our-programs/food-and-farms

Local Harvest   localharvest.org

Rodale Institute   rodaleinstitute.org

Slow Food USA   slowfoodusa.org

Sustainable Table   sustainabletable.org

INDEX

Lisa D. Holmes is a Portland, Oregon, designer and author. She grew up in Kansas City, Missouri, where she attended the Kansas City Art Institute, majoring in graphic design. Lisa has devoted the majority of her career to graphic design, working in both traditional print and digital media. She has worked as a designer and art director for both design studios and corporate entities, as well as her own communications design firm, Yulan Studio.

Her work in various arts and crafts is highly eclectic and has included ceramics, drawing and painting, photography, and textiles. She knits, crochets, and designs and sews her own clothing. Lisa also is the author of *I Heart Oregon (& Washington): 25 of the Portland Area's Best Hikes*. She is currently at work on a follow-up book, which she hopes will keep her exploring the beautiful environments of her beloved Pacific Northwest. You can follow her adventures on her blog: iheartpacificnorthwest.com.

John Clark Vincent is a writer and author who lives in Portland, Oregon, with his wife, Lisa D. Holmes, and their dogs Sadie and Cricket. He grew up on a family farm in central Kansas, then attended the University of Kansas where he studied creative writing and language arts education. While at KU, he was accepted into the English Honors Program and was awarded first place in the William Herbert Carruth Memorial Poetry Contest. He has spent the majority of his career working as a writer and producer in marketing communications, including fifteen years at Yulan Studio, which he owns with his wife, Lisa.

In addition to *Planting A Future: Portraits from Oregon's New Farm Movement*, John co-authored *Winemakers of the Willamette Valley: Pioneering Vintners from Oregon's Wine Country*, and published a collection of poetry, *Repairing Shattered Glass*. John's poetry and screenplays have won awards from the Willamette Writers' annual Kay Snow Writing Contest.

John's interest in agriculture began as a child in Rice County, Kansas, where he worked on his family's farm. The dryland farm produced wheat, various sorghums, alfalfa, and occasionally crops like sudangrass or cow peas for fodder and soil improvement. Livestock included cattle and hogs for commercial production, chickens for family eggs, and a few lambs for 4-H projects. His 4-H experiences led him to a fascination with entomology, and his informal study of insects continues today. He is an avid gardener and loves to ferment his own produce.